I Want to be a
Product Manager
From developer to Product Manager

by
Claudio Marrero

Javascript

console.log("Hello World")

Python

print("Hello World")

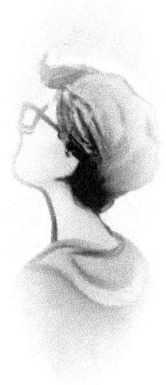

ÍNDICE

c#

```
Console.WriteLine("Hello World")
```

DEDICATION

I dedicate this book to my beloved wife, who has been my rock throughout this entire process. Her love, support, and encouragement have been crucial in enabling me to complete this project.

This book is a small testament to all that we have built together.

I love you.

PHP

```
echo "Hello World"
```

PROLOGUE

I n 2013, while leading a start-up as a CTO, I had a revelation after our initial launch turned into a failure: the value perceived by the user is often less than the real value proposed. This, of course, is not new to any product owner or product manager who is dedicated to receiving user feedback, it's the norm.

This "discovery" led me to change careers and focus on understanding the user, their real problem, how they perceive it, and creating strategies based on that problem, rather than focusing on creating technologically perfect products with thousands of amazing features. In other words, I changed my approach from "Hello World" to "Hello User".

This book is mainly aimed at people who have a technical present or past, be it in programming, design, or any other specialization within the value chain of technological product development, but who want to redirect their career towards the role of product manager. However, I hope it will also be highly useful for product owners or product managers with a technical background who want to take their career to the next level.

This is an increasingly common change, as it's a logical transition. The good news is that experience in almost any technical specialization can be extremely valuable in the field of product management, as both roles involve problem-solving and logical thinking. However, it's important to note that a different set of skills and knowledge is required compared to what's needed to be a good specialist.

In this book, I'll not only tell you everything you need to know about how to perform as a product manager but also about all the skills that no one tells you that you need to have, how to develop them, and what you should focus on depending on the stage of your career.

This book is designed by levels, just like when you start a game you have a tutorial where you learn the basic techniques, concepts, how to interact with each element, etc., and then you progress through each level, experimenting and acquiring new skills, evolving into or towards a new role requires learning the basics and then honing each of the learned techniques. Thus, I hope this book will accompany you throughout your career and that you'll be able to interpret from different points of view the challenges I'll present to you throughout it.

Just like with any product, a good product manager must gather feedback for absolutely everything, which is why the opinion of over 156 product managers, each with a completely different story and experience, has been fundamental for me. They helped me to objectify my vision and purpose with this book. For this reason, you'll find an interesting mix of opinions and views from experienced professionals in different areas throughout each page, which made the task of translating the complex world of product management into simple examples and explanations easier.

Without further ado, let's get started!

Java

```
System.out.println("Hello World");
```

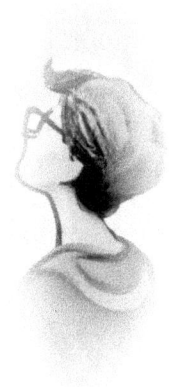

GETTING STARTED

INTRODUCTION

I t's difficult to attribute the invention of the product manager role to a single person or company, as this kind of position has evolved over time in response to the needs of businesses and market changes. However, Neil H. McElroy, President of Procter & Gamble from 1957 to 1960, is generally considered one of the pioneers in brand management and the creation of teams dedicated to managing individual products. McElroy wrote an internal memo in 1957 discussing the importance of focusing on the development and promotion of individual brands, rather than treating all company products uniformly. This led to the creation of a brand management system where each product was assigned to a team of product managers responsible for its development and promotion. This approach has become a common practice in many companies and is considered one of McElroy's main contributions.

Though it may be difficult to attribute the invention of the product manager role to a single person or company, when we observe the evolution of the role over time, we can see a pattern: the increase in the diversification of specializations and responsibilities. What does this mean? Well, rather than having one general manager for all brands, diversifying focus by putting different people in charge of each one allows each company to better analyze the changes that need to be made to each product and, in this way, make decisions more quickly, potentially saving thousands of dollars in the process.

Nowadays, it's common to see tech companies dividing the product manager role into more specific specializations. For instance, in an e-commerce business, you might have a product manager (PM) focusing on the user experience while navigating the site and viewing the different products on offer, and another PM focusing on the checkout or shopping cart section. This allows the company to more clearly and specifically translate the actions the team needs to undertake to evolve the product.

This role is precisely **the intersection between specialization and the management and understanding of people.**

WHAT IS PRODUCT MANAGEMENT?

In its purest form, the role is the process of guiding the development and implementation of a product, from strategy definition to launch and management of the product's lifecycle. This includes working with an interdisciplinary team of engineers, designers, marketers, and other professionals to ensure that the product meets customers' needs and expectations and contributes to the company's long-term success. Product management involves a user-centered approach and constant iteration and improvement based on data and feedback.

But in practice, if we analyze my intersection definition, the PM is a translator of the needs of users and stakeholders (those interested in the results obtained from your management) into a strategic language that can be understood by both the company's management and the entire team of specialists, be they programmers, marketers, designers, product owners, product analysts, etc.

The complex decisions of a PM vary from company to company, and even more from sector to sector, due to the fact that it's a role that must adapt to each product. Hence, the responsibilities, experiences, and knowledge of a PM can vary significantly. If we take a company that manufactures cars and another very different one, such as a streaming platform, the responsibilities, knowledge, challenges, and daily tasks will be extremes of the same role. That's why if you are on this journey, you must understand that the sector in which you work will greatly guide your career.

Those who usually excel in this role are those who have a deep understanding of the sector in which their company operates, who are passionate about solving problems in that sector, and who strive to learn even more about how that market interacts with their product and other similar products.

In the remainder of this beginner's tutorial, we'll attempt to take a look at all the techniques, methodologies, and best practices of this position, and even if you forget them in the very short term, they'll be of great help when we explore the more advanced levels.

WHAT ARE THEIR RESPONSIBILITIES AND CHALLENGES?

We'll be responsible for defining the product's strategy and focus, engaging customers and gathering their feedback, working with the development team and marketing team to launch successful products, measuring their success, and making adjustments accordingly, leading a product team, and fostering a culture of innovation.

Some of the challenges we'll face include balancing the needs of customers with those of the company, ensuring the product is launched on time and within budget, and dealing with constant changes in the market and technology. Additionally, we'll need to be able to adapt quickly and make decisions based on incomplete and changing information.

When I started navigating these waters, companies were still uncertain about the effectiveness of incorporating such roles into their architectures, which is why there was little information available. Today it's very easy to get lost in the plethora of blogs, books, and courses that exist online, and it's common to find information that tries to sell the role without detailing the responsibilities that a PM must assume. Therefore, I've attempted to summarize them in these twelve concrete points: prioritization, teamwork, time management, effective communication, decision-making, market changes, scalability, balancing innovation and profitability, change management, expectation management, risk management, and quality management. Depending on which techniques, methodologies, or practices you decide to use, you'll have one management style or another, but none can guarantee success, as we'll see later on.

PRIORITIZATION

One of the primary challenges for PMs is deciding which features or functionalities should be included in the product. This is particularly difficult when there are a lot of suggestions and requirements from different stakeholders.

Task prioritization begins with a clear understanding of long-term objectives. Once these goals have been set, it is possible to work backward to identify the critical tasks that need to be performed to achieve them; in some methodologies, this is called retro-planning. These tasks must be prioritized based on their importance and urgency, and the expected impact they will have on long-term goals should be considered.

It is necessary to be realistic about how much can be achieved in a given period and assign resources to the tasks that have the most significant impact. This means that you may have to delegate less important tasks to other team members or postpone them until a more suitable time.

Another key factor in task prioritization is maintaining a focus on flexibility and adaptability. In the world of product management, things change quickly, and new tasks or unexpected problems may arise. You'll need to be able to constantly reassess and realign tasks to ensure that you are always focusing on the most important and relevant tasks.

There are countless methods, frameworks, and methodologies for prioritizing these activities; although we will delve deeper into this later on, here are some of the most well-known ones:

MOSCOW Method

This method divides tasks into four categories: Must-have, Should-have, Could-have, and Won't-have at this moment. This allows teams to prioritize tasks and focus on the most important ones first.

I specifically use this methodology to organize my own tasks and stay up-to-date with the most important things, but I do not transmit or coordinate using it, as it has not given me good results for teamwork, especially if it involves large teams.

KANBAN Method

This method uses a visual board to track the progress of tasks through different stages of development. Teams can use this method to identify and prioritize the most important tasks and ensure they are advancing effectively.

Tools such as JIRA, Trello, Asana, Microsoft Planner, and almost all management tools have a Kanban board, and while it is very useful for organizing the beginning of a project, it quickly becomes complex to manage and follow up. I only recommend it for short iteration fragments, like sprints or design thinking sessions, among others.

ICE Method

This system evaluates tasks in terms of impact, confidence, and effort. Teams can use this evaluation to prioritize the most important tasks and ensure they are focusing on those that have the greatest impact on the product and the company.

RICE Method

This is similar to the ICE method, but it also includes an assessment of the implementation speed. Teams can use this method to prioritize the most important tasks and ensure they are focusing on those that can be implemented in the fastest and most effective way.

Both the ICE and RICE methods are very effective when your team needs to deal with BaU (Business as Usual). Ordering the backlog of tasks with these methodologies is highly effective, but it brings the complexity of analyzing each task, which can generate a lot of controversy when it comes to assigning a value to each pending issue.

PICK Method

This allows teams to select a limited number of tasks to work on each development cycle. It enables teams to focus on fewer tasks to ensure they are advancing effectively.

Value and Effort Method

This evaluates tasks in terms of their value to the business and the effort required to complete them. Teams can use this evaluation to prioritize the most important tasks and ensure that they are focusing on those that have the greatest impact on the product and the company.

Value and Deadline Method

Similar to the value and effort method, it also includes an assessment of the deadline to complete the task. This can be used to prioritize the most important tasks and ensure that they are focusing on those that need to be completed in a shorter period of time.

Value and Risk Method

This is similar to the value and effort method but also includes an assessment of the risk associated with each task. It can be used to prioritize the most important tasks and ensure that they are focusing on those that carry the least risk of non-compliance.

In conclusion, prioritize according to the results that you and your stakeholders want to obtain. There is no perfect methodology, each one is designed for a type of company, for a work style, and for different objectives.

In my case, I don't use a single methodology. I take the best from the ones I mentioned before, and depending on the team, the expectations of the stakeholders, the market I'm targeting, the maturity of the product (i.e., whether it has just started or has been on the market for several years), I use one methodology or another. However, I always maintain transparency towards the entire team about what the prioritization rules and the reasons are, which helps keep them motivated and fully in context when changes are very drastic.

TEAMWORK

A great deal of technical and business skills is required to ensure that a product is a success in the market. As a programmer, you probably already have a strong understanding of the technical aspects of product development. However, success also relies on skills such as communication, negotiation, and leadership ability.

Teamwork is essential in ensuring all these skills effectively combine to achieve a successful end result. By working with other departments, such as marketing, finance, sales, etc., a more comprehensive and balanced view of market and customer needs can be obtained. This in turn allows for informed decisions about product development and strategy.

Therefore, the PM must work closely with other departments, and for this, there are various methodologies that you can use to help you be effective in this process:

Scrum

This agile methodology focuses on rapid delivery of small work iterations. Scrum teams are divided into small multidisciplinary groups and use short daily meetings called "stand-ups" to stay up-to-date and solve problems.

This methodology is, today, probably the most well-known and the one most software companies use, but don't hesitate to adapt it according to the maturity of your teams; the process can be very exhausting for those who haven't used it in a purist way.

Lean

This focuses on avoiding waste or time losses and rapid delivery. Lean teams use techniques such as "value analysis" to identify and eliminate activities that do not add value to the product.

Design thinking

This centers the approach on the user and solving problems creatively. Teams use techniques like empathy and rapid prototyping to understand user needs and develop innovative solutions. Using design thinking can help you work more effectively with design and engineering teams to develop products that meet user needs.

Safe (*scaled agile framework*)

This is a project management methodology designed to speed up the delivery of software and services. It uses a waterfall and iteration-based approach to plan and deliver work incrementally and continuously. This methodology is designed to scale to teams and projects of different sizes and focuses on the rapid and continuous delivery of value to the customer. In SAFe, projects are broken down into small parts called "iterations" and a Kanban board is used to visualize ongoing work and progress. Agile planning techniques, such as user story planning and value stream mapping, are also used to define and prioritize the work that needs to be done.

TIME MANAGEMENT

Being efficient in managing your time and ensuring your team is as well is one of the main responsibilities you'll have as a product manager. You probably already have a deep understanding of the importance of efficiency and organization in software development. However, in product management, time management becomes even more important due to the number of responsibilities and tasks that need to be addressed to ensure the success of a product.

There's a lot of controversy regarding team time management and many misunderstandings. For example, the fact that the team works result-oriented doesn't mean they shouldn't manage time or that there shouldn't be roadmaps or deadlines for tasks. Administrative tasks tend to be very tedious for the team, and the chances of making mistakes in estimates in a rapidly changing market are very high. Fortunately, we have some methodologies or good practices that help us in this management:

Gtd (*getting things done*)

This method focuses on organizing and prioritizing tasks. It's based on the idea of capturing all tasks and projects on a central list and then organizing them in order of priority to maximize efficiency and performance.

I highly recommend you to read David Allen's book, "Getting Things Done," where he talks in detail about this methodology, which is more of a productivity and time management system that helps you complete tasks and manage them.

Timeboxing

This method focuses on assigning specific blocks of time to each task or project. Product managers can use timeboxing to ensure they are devoting the right amount of time to each task and to avoid distractions and deviations from their work plan.

During my transition from programming to product management, passing through roles such as technical lead, CTO, etc., I had to personally organize myself in this way. I created an Excel where I categorized the general tasks of my day-to-day with the time I dedicated to each one, and at the end of the day, I could see where I was focusing my time and discover why I couldn't finish the most important tasks as planned.

Lean time management

This method focuses on the elimination of time waste and its optimization. Product managers can use techniques like "value analysis" and "elimination of unnecessary tasks" to ensure they're focusing their time on the most important tasks and activities. Moreover, techniques like "batch planning" and "branch pruning" can be applied to ensure the team is not overtasked and is working effectively. Using Lean Time Management can make product managers more productive and achieve their goals more efficiently.

Pomodoro

This method is based on the idea of working in 25-minute blocks of time, followed by a short break. The idea is to stay focused on a single task during the work block and then take a brief break to recharge and prepare for the next work block.

While not strictly a product time management methodology, I've decided to mention this practice because it's incredibly useful when calculating that each team member achieves approximately a day and a half of rest per month if you take into account the 15-minute breaks for every hour of work.

Also, it's a practice I highly recommend to maintain team health and productivity, and it can be implemented alongside any other management methodology you decide to implement.

COMMUNICATION

You must be able to effectively communicate your product vision and strategy to various audiences, including employees, investors, and customers, among others.

Bernard Werber published in his trilogy "The Ants" this phrase that I find exceptional: "Between what I think, what I want to say, what I believe I say, what I say, what you want to hear, what you believe you hear, what you hear, what you want to understand, what you believe to understand, there are ten possibilities that there will be problems in communication," and if you've noticed, there are only nine.

We must assume that communication problems will be inevitable, regardless of the methodology you use; all you can do is reduce the chances of creating misunderstandings, and for that, as with the rest of the points, I can give you some best practices:

Establish a clear and open line of communication

Product managers must establish a clear and open line of communication with their team and other key stakeholders. This can include regular meetings, written updates, and using online collaboration tools to keep everyone informed about what is happening.

Work on problem-solving

When problems or disagreements arise, you must work with others to find solutions and reach concrete agreements. More than once you will have to make concessions and it will be extremely uncomfortable, but never let a detected problem mature, be it internal or external.

Use Different Communication Channels

It's crucial to ensure you reach all relevant individuals. This can involve face-to-face meetings, emails, text messages, phone calls, and online collaboration tools. Consider the most appropriate communication medium for each situation and make sure to use the one that effectively conveys information in a clear and concise manner.

DECISION MAKING

One of the most significant challenges you will face is making difficult decisions and remaining calm in high-stress situations. Often, you'll need to do this based on incomplete or conflicting data, but it's important that you can carry it out swiftly and reliably.

From my perspective, the worst decision is not to make any; if there's one thing we need to do, it's continuous learning, and the most effective way to do that is by making mistakes. Hence, if we don't make decisions, we won't have the opportunity to learn from our mistakes.

The fundamental thing is to make "cheap" mistakes, and the following methodologies can help with that:

Lean Thinking

This is based on eliminating time-wasting tasks and optimizing value. You can use techniques like "value analysis" and "elimination of unnecessary tasks" to ensure you're focusing your time and effort on the most important tasks and activities.

Agile Decision Making

This involves quick adaptation and constant iteration. You can use techniques like "rapid prototyping" and "constant feedback" to ensure you're making decisions quickly and effectively and that you're adapting to market changes and user needs.

Data Analysis

This can be a powerful tool for making informed decisions. Product managers can employ techniques such as "key metric analysis" and "customer survey analysis" to gain a deep understanding of what is working and what is not, and thus make decisions based on solid data.

Principle-based Decision Making

This is grounded in key values and principles. Product managers can use this technique to ensure their decisions align with the company's values and goals, avoiding decisions that could jeopardize its ethics or integrity. Some questions product managers can ask themselves when using this technique are: Is this decision consistent with our key values and principles? Is this decision beneficial for our customers and end-users? Is it beneficial for the company in the long term?

A reference you can read about principle-based decision making is Ray Dalio. His book, titled Principles, is one of my favorites, and I try to reread it whenever I can.

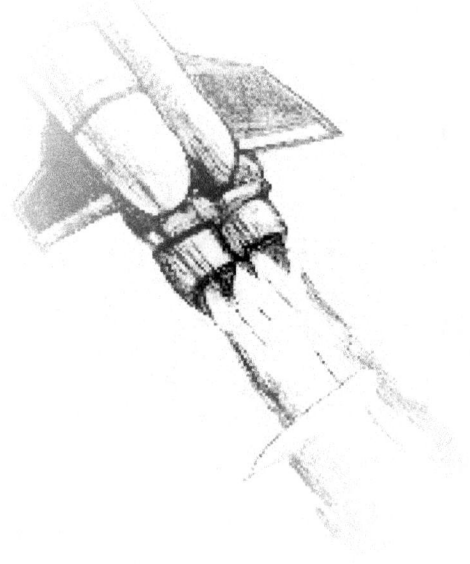

MARKET CHANGES

You must always be alert to changes in the market and adapt your strategy accordingly. This may involve introducing new products or modifying existing ones to meet customers' changing needs.

There are many well-known examples of companies that failed to do so, like Blockbuster or Kodak, to mention some of the most popular and significant failures. Besides the techniques I'll mention in this point, one activity I strongly recommend is keeping up-to-date with the birth of new startups, which ones have been invested in or acquired, which market they're targeting, and how they relate to your product. On more than one occasion, this has helped me make the right decision to change strategy and ensure competitiveness..

Conduct Market Research

This involves keeping abreast of trends and changes in your company's market. This could include conducting market surveys, participating in events and conferences, and reading reports and studies.

Engage in Professional Networks

Product managers can greatly benefit from participating in professional networks and attending industry events and conferences. This allows them to connect with other professionals and stay up-to-date on the latest trends and developments in the sector.

Conduct Feedback and Follow-Up Sessions

We'll talk a lot about this throughout the book, as regular feedback and follow-up sessions with your teams and other departments in the company are fundamental to the product development lifecycle. This allows product managers to keep track of changes and make informed decisions.

Use Tracking and Analysis Tools

There are many tracking and analysis tools available that can help PMs stay on top of changes and trends in the market. For example, for an e-commerce, it will be useful to have tools for monitoring competitors' prices, and for newspapers or magazines, knowing the most read categories of their rivals. This will influence the decisions you need to make when prioritizing your strategies.

By integrating all the internal information about your product with what you can obtain from the market and your competition, you can make decisions with a lower probability of failure and ensure a growth rate with moderate risk.

SCALABILITY

As the product grows, the product manager must have a plan to scale and manage growth sustainably. This includes hiring and managing a larger team, and implementing appropriate processes and systems to support growth.

One can fail by doing things poorly, but also by doing them well. The inability to handle a substantial and sustained increase in sales, that is, managing the responsibility of being successful, is as important or more so when faced with failure.

Lean Startup Approach

While I've already mentioned the Lean methodology several times, I'll continue to refer to this approach as Lean Startup focuses on creating Minimum Viable Products (MVPs) and performing rapid, ongoing market tests to see how the public responds. This allows product managers to make informed decisions on how to scale the product and make changes quickly, based on market feedback.

Growth Hacking Techniques

This focuses on using unconventional techniques and tools to drive product growth. This can include using digital marketing techniques, automating processes to drive product growth, and implementing PLG (product-led growth) strategies, among others.

Business Model Generation Techniques

This focuses on creating a solid business model and constantly reviewing and improving that model. This can be useful for product managers when trying to scale products and ensuring that the business model is sustainable in the long term.

Alexander Osterwalder has written an exceptional book on how to create these models, Business Model Generation. Its reading is not recommended, but mandatory if you are or want to lead a product team.

Customer Development Techniques

This is based on learning through direct contact with customers and obtaining continuous feedback. Product managers will use this process in each iteration when trying to scale products and ensuring they are meeting customers' needs.

While it is part of the Lean methodology, it deserves to be discussed separately, as we will use these techniques individually to obtain information about the permanent changes in our market.

BALANCE BETWEEN INNOVATION AND PROFITABILITY

Balancing innovation and profitability is one of the most significant challenges a product manager faces. On the one hand, innovation is crucial to staying competitive and attracting customers, but on the other hand, it's essential to ensure that the product is profitable and yields a positive return on investment.

The product manager must strike a balance between these two factors when evaluating and prioritizing product features and functionalities. This can involve considering factors such as development cost, revenue potential, and ability to differentiate in the market.

This means making sure that investments are being made in the product in areas that genuinely matter to customers and that an adequate return is being obtained on those investments.

Managing profitable innovation should be a process included in each of the methodologies proposed when setting goals, prioritizing them, and executing them, and in updating technologies, creating new features, launching new products.

While we will explore in detail in this book the different strategies, it's important to mention some very specific ones:

Focus on Value

Develop features and functions that truly add value for customers and differentiate them from the competition.

Prioritization

Prioritize product features and functions based on their impact on profitability and customer satisfaction.

Leverage Emerging Technologies

Stay up-to-date with emerging technologies and explore how they can be used to enhance the product and increase its profitability..

Fast Prototyping

Develop quick prototypes to test and validate innovative ideas before investing in their large-scale development.

If you include a percentage of innovation in everything you do every day, you should never have to worry about it.

CHANGE MANAGEMENT

You must be able to lead change and convince others of the need to adopt new strategies and approaches. This can be particularly challenging when dealing with significant changes that affect the company's culture or the way things are done.

The great challenge of "change" is the fear generated by the unknown of what is on the other side of the change, and these fears are generally associated with a culture that penalizes failure.

Transformational Leadership

An approach in which the leader focuses on inspiring and motivating employees through a clear vision and communication of goals.

An effective transformational leader understands the importance of involving their team in the change process. This means listening to the ideas and opinions of team members and working together to find innovative solutions. By doing this, the leader can ensure that all stakeholders are aligned and working together toward a common goal.

Situational Leadership

It focuses on adapting the leadership style to the situation and the team's needs, rather than using a single leadership style for all situations.

Situational leadership involves three key components: self-awareness, referring to understanding one's strengths and weaknesses as a leader, knowledge of team members, which encompasses understanding the needs, skills, and motivations of each team member, and adaptability, changing the leadership style based on the specific needs and circumstances of the situation.

Coaching

It involves using coaching techniques to help employees develop necessary skills and competencies to adapt to change.

It is a process of support and personal development that helps people achieve their goals and reach their full potential. In the context of change management, coaching can be a valuable tool to accompany individuals and teams through processes of change and transformation.

It can help leaders and teams navigate the change process more effectively and provide a safe and confidential space for people to explore their thoughts, feelings, and challenges related to change. Using coaching techniques, you can help individuals identify and address barriers and challenges, and develop skills and strategies to effectively handle change.

In the context of a business reorganization, it's common for leaders and teams to feel uncertainty, anxiety, and stress related to change. A coach can work with these leaders and teams to help them identify and address their emotional challenges and develop strategies to navigate the change process effectively.

Gradual Change

Implementing change gradually and progressively, rather than suddenly, can help employees feel more comfortable and better adapt to change.

It's often a preferred strategy to implement changes in complex and regulated business environments, where radical change can have unpredictable and negative consequences. Gradual change is also an effective strategy to implement changes in sensitive areas, like culture and labor relations, where radical change can be harmful.

However, gradual change can also be slower and more expensive than radical change, and it may require more planning and coordination to ensure the changes are effectively and sustainably implemented.

Management of Fear of Change

Recognizing and effectively addressing employees' fear of change can help increase their acceptance of it. It is a natural response to uncertainty and threat, and can manifest as resistance, anxiety, negativity, or even sabotage.

Training and Development

Providing training and development to help employees acquire the necessary skills and competencies to adapt to change. By investing in employee training and development, companies can improve productivity, increase employee motivation and satisfaction, and prepare for future challenges.

Ensure the Support and Participation of Top Management

Securing the support and participation of senior management is crucial to ensuring the success of long-term change.

Top management is responsible for setting the organization's vision and objectives, and for making major strategic decisions. Without this support, implementation is likely to be much more difficult or directly impossible as they are the ones who provide the financial, human, and political resources that are essential for the success of change.

Securing the support and participation of top management is not something that can be achieved overnight. It's an ongoing process that requires clear communication, a well-thought-out strategy, and an effective action plan. Organizational leaders should be informed about the objectives and risks of change, and should be aware of their role in its success.

How many times have you proposed new technologies, features, methodologies, etc. that were rejected both by your direct boss and the rest of the company's executives? In my case, I can't count them all with my fingers and that's because there's a hidden cost to align all necessary parties for change to be effective, and this is not always easy to see and provide a clear strategy.

Planning and Strategy

Use planning and strategy to anticipate and prepare for change, which can help minimize its negative impact.

Planning is critical to the success of change management, as it allows organizational leaders to carefully consider the necessary resources, timings, risks, and impacts of change before implementing it. It also allows for identifying any challenge that may arise during the implementation process, and to develop proactive solutions to address them.

The strategy, on the other hand, is the overall action plan used to achieve the objectives set out in the planning. The strategy should be clear, flexible, and adaptable to changes, as change management is often a dynamic process that requires adjustments as it's implemented.

Effective Communication

Use effective communication to ensure that all employees are aware of changes and understand the reason behind them. However, remember that it's more than just transmitting information. It also includes actively listening and considering the perspectives of all stakeholders. This allows organizational leaders to address any concerns or challenges that arise during the implementation process.

Messages should be easily understandable to all recipients and should be conveyed consistently and uniformly.

Communication should also be two-way, meaning there should be an exchange of information and feedback between organizational leaders and employees. This allows organizational leaders to adjust their approach and address any challenge that arises as organizational changes are implemented.

MANAGEMENT OF EXPECTATIONS

You must be able to manage the expectations of different stakeholders, including employees, investors, and customers. This means ensuring that everyone understands the long-term objectives and how the product is approaching them, as well as appropriately managing expectations in terms of timelines and features.

Of all the points mentioned so far, I believe this particular one is one of the most complex, which is why I would like to quote both **Eric Ries** and **Steve Blank**, both experts in Lean methodologies:

Staged lifecycle or *build-measure-learn*

Ries introduces this approach in his book The Lean Startup, where he emphasizes the importance of building a Minimum Viable Product (MVP) and measuring the results at each stage to learn and continually improve. This allows product teams to manage the expectations of investors and other stakeholders by providing a clear view of the progress made and how they are approaching long-term goals

Cascading goals or waterfall experiments

Blank uses this approach in his book The Four Steps to the Epiphany to help companies develop products more efficiently. It is based on the idea that goals should be set at the company level, and then broken down into more specific goals at the product and team level. This allows product teams to manage expectations more effectively by providing a clear structure to achieve long-term objectives.

Customer development methodology

Ries and Blank have also highlighted the importance of involving customers in the product development process. This can include conducting customer interviews, designing surveys, and participating in focus groups to better understand the needs and expectations of buyers. By involving customers in this way, product teams can better manage buyer expectations by providing a product that meets their needs.

RISK MANAGEMENT

Evaluating and managing the risks associated with launching and developing a product is an ongoing challenge that you will need to consider. This may include identifying potential technical or market problems and implementing measures to mitigate those risks.

Risk Analysis

This technique involves recognizing and evaluating potential risks associated with a project. This includes identifying factors that could negatively impact the project, assessing the likelihood of these risks occurring, and determining the consequences if they do occur.

Contingency Planning

This involves developing action plans to address identified threats. These actions include implementing preventive measures to avoid risks from occurring and response plans to address problems if they occur.

Risk Monitoring and Control

Once measures have been implemented to manage risks, it's important to continuously monitor the project's progress to detect any changes in the risk landscape. This can include conducting regular risk assessments and implementing control measures to mitigate any new or changing threats.

QUALITY MANAGEMENT

Quality management is a comprehensive approach to continually improving an organization's products and services' effectiveness and efficiency. It is about ensuring that products and services meet customers' expectations and industry quality standards. You will be responsible for ensuring that the product meets quality standards and customer requirements.

This includes managing quality during product development and launch, as well as resolving quality issues once the product is on the market.

Quality in Design

This involves ensuring product quality from the outset, through identifying and eliminating potential failures during the design phase. Of course, the product manager is not the one conducting these tests, but they must ensure they are carried out conscientiously and dedicate the appropriate time to the project, including quality management in each cycle.

Monitoring and Measuring Quality

This involves establishing metrics and monitoring the product's performance to ensure it meets quality standards. This involves collecting and analyzing data on the quality of products and services. The results of these analyses should be used to identify areas for improvement and develop strategies to enhance quality.

Team Training and Development

This involves ensuring that the team has the necessary skills and knowledge to work effectively in quality management.

It's a challenging role that requires a combination of skills, techniques, and leadership. You must be able to work as a team, make difficult decisions, manage the time and expectations of various stakeholders, and adapt to changes in the market.

You will also need to guarantee product quality and manage the risks associated with its development and launch.

¿PROJECT MANAGER OR PRODUCT MANAGER?

Nowadays, both company executives and HR staff responsible for personnel search and selection no longer tend to make these mistakes. However, it was common to read job offers with the title of product manager and a list of responsibilities and challenges that literally define a project manager.

On more than one occasion, I have had to explain the difference to various executives of highly recognized companies, with the aid of slides and amidst many skeptical looks. However, these same executives currently cannot conceive of a business structure without this role.

If I had to choose what seems most important to me about their differences, it is that a project manager does not need to have knowledge of a market sector to perfectly fulfill his role. This is because it involves the management of tasks, their delivery dates, and the follow-up but not the definition and prioritization of tasks based on the needs of a target market.

PROJECT MANAGEMENT & PRODUCT MANAGEMENT

Project Management	Product Management
Focuses on the planning, organization, and execution of short-term projects, with defined scope and objectives	Focuses on long-term strategy and the creation and development of products
Centers on short-term results delivery and meeting set times and budgets	Centers on the long-term value that the product brings to customers and the company
Deals with resource management and coordination of interdisciplinary teams to carry out specific projects	Deals with product lifecycle management and strategic decision-making regarding its direction
Tends to be a more tactical role focusing on task execution and short-term problem-solving	Tends to be a more strategic role focusing on long-term decision-making regarding product direction and identification of market opportunities

In terms of the typical growth of each position, the project manager often requires more technical and operational skills, such as the ability to efficiently plan and organize projects, work under pressure, and with tight deadlines. The product manager, on the other hand, often requires a combination of technical and strategic skills, such as understanding customer and market needs, making long-term decisions about product direction, and leading development and marketing teams.

In terms of academic training, people who want to be project managers commonly have an educational background in engineering, computer science, or information science; while those who aspire to be product managers may have a background in any field, from engineering to business or marketing.

In terms of career advancement, both roles tend to have high growth potential and can lead to leadership positions in the technology or business area. However, project management often focuses on short-term delivery of specific projects, while product management focuses on long-term strategy and decision-making about product direction. Therefore, individuals who want to advance in the field of product management may have greater opportunities to lead teams and make strategic company-wide decisions.

It's very common for start-up founders, company CEOs, and business leaders in general to have worked as product managers or similar roles.

ROLES THAT USUALLY RELATE TO A PRODUCT MANAGE

Each company has a unique structure, both the names of the positions and the responsibility and focus of each one is different in each business, but in general terms, you will be in constant relation with the following roles:

Head of product

This is the leader of the product team and generally your direct leader. They are in charge of long-term strategy and decision-making regarding product direction. Moreover, they manage the product life cycle and ensure that the product team is aligned with the company's strategy.

Product owner

This is who a PM directly leads, and their main ally throughout the product development process. They are responsible for defining the product vision and purpose, and setting goals and success metrics for it. They are also responsible for engaging customers and getting their feedback, and working with the development team to ensure deadlines and budgets are met.

Scrum master

This person ensures that the development team follows the Agile Scrum framework and removes obstacles that prevent team progress. They are also responsible for facilitating teamwork and ensuring that the team follows the product development process.

Technical Lead

This is the person in charge of leading and coordinating a team of engineers and developers to meet the product's technical objectives and quality. They must have exhaustive knowledge of relevant technologies and platforms and be able to make important technical decisions.

UX

They focus on how the user feels when interacting with a product. Their role is crucial in ensuring the product is easy to use and attractive to users.

It is essential to understand the company's goals and objectives and how the product contributes to them, to understand the users, both internal and external, and how the product can address their needs and problems.

I do not want to forget to mention the importance of identifying and managing conflicts and personal agendas that could affect the product creation process.

Being a successful product manager requires a combination of technical and strategic skills, as well as a deep understanding of the needs and challenges of all parties involved.

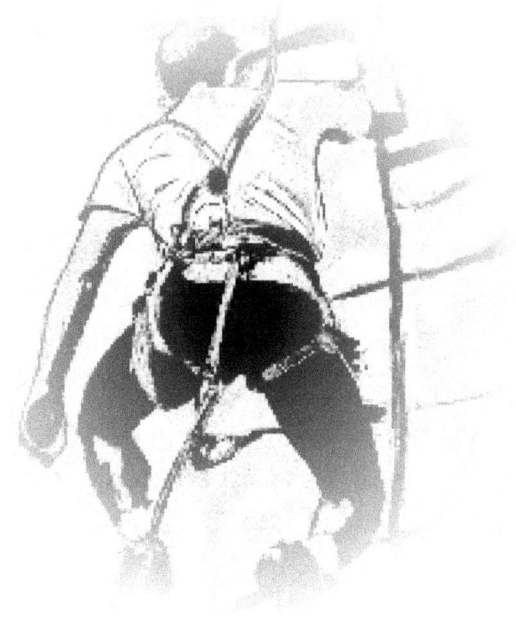

WHAT QUALITIES ARE NECESSARY TO BE A PRODUCT MANAGER?

At the same time as I began to manage people, I started to develop many of the skills I later needed as a PM; through mistakes and thousands of different attempts, I understood what motivated my teams and the leadership style I wanted to have.

While different methodologies and frameworks allow you to be organized, methodical and reduce the amount of errors that are usually made, I also discovered that you can't lead two different people in the same way, and what truly matters is to be flexible and highly creative to figure out how to make your team meet its goals, without diminishing their motivation and performance. This has been different with each group: I had teams that required an excess of communication through multiple channels and a constant repetition of goals, and others that needed to meet only once a week, at most.

The only constant is the ability to adapt to the team's dynamics.

So, **WHAT** SKILLS MUST YOU DEVELOP TO FULFILL THIS ROLE?

Just like I divided the responsibilities and challenges into twelve key points, there are four skills you should develop: technical understanding, communication skills, leadership and management skills, and analytical and strategic thinking.

TECHNICAL UNDERSTANDING

Coming from the world of programming has given me enormous advantages when making decisions. Understanding the technological world from its base, debating with different teams about technological proposals and, above all, understanding them in detail allows you to contrast them with the needs of the company and your customers; creating that map from one end to the other is, for me, the most important skill. In fact, that's why I'm writing this book, and I invite all those specialists, whether they are programmers, designers, etc., to venture into the world of product development from a more strategic and not so operational perspective.

While you don't need to be an expert in programming, design, or another specialty, it's important to have a solid foundation in technological knowledge and understand how products work. This can be especially useful when working with the development team to ensure deadlines and budgets are met.

COMMUNICATION SKILLS

Meetings that last more than fifteen minutes, constant defocusing from what's important, egotistical interlocutors who love to hear themselves talk, presentations with endless texts... We've all been there at some point, and that's why it's essential to develop the ability to understand why, how, when, and with whom you should communicate.

You will need to be able to express yourself clearly and effectively with both the development and marketing teams. This includes presentation, writing, and negotiation skills, among others.

Active Listening Skills

You must be able to attentively listen to your colleagues, customers, and other stakeholders to understand their needs and expectations.

What does active listening mean?

- Giving full attention to the speaker, maintaining eye contact, and not getting distracted by other issues.

- Asking open-ended questions helps delve into the topic and gather more information.

- Summarizing and reflecting on what has been said ensures you've understood correctly and provides a space for reflection to show you've been attentive.

- Avoid interrupting and let the speaker finish expressing themselves before asking questions or adding comments.

- Showing empathy to demonstrate understanding of the speaker's emotions and perspectives establishes a connection and creates a trustful environment.

Presentation Skills

Being capable of presenting the vision and strategy clearly and concisely, both to the team and other stakeholders.

- Dedicate the necessary time to adequately prepare the presentation, including its structure, content, and visual aids.

- Using a clear, firm voice and confident body language helps capture the audience's attention and effectively transmit the message.

- Using graphs, images, and other visual aids helps illustrate the message and makes the presentation more appealing and easy to understand.

- Ensure the presentation fits the available time and doesn't extend too long.

- Establish a dialogue with the audience and respond to questions and comments clearly and concisely.

Negotiation Skills

You'll have to negotiate with different departments and stakeholders to get the support and resources needed to meet your goals. In my opinion, this is one of the most complex and hard-to-learn skills, full of compromises and frustrations. Some negotiation techniques include:

- **Interest-based negotiation:** Instead of focusing on rigid positions, this technique focuses on understanding the underlying needs and interests of each party and seeks solutions that benefit everyone.

- **Win-win negotiation:** This is based on finding solutions that are beneficial to both parties, instead of trying to impose a position.

- **Distributive negotiation:** Also known as "win-lose", this technique aims to get the most possible benefit for one party at the expense of the other.

- **Collaborative negotiation:** This is grounded on the joint work of the parties to find creative and beneficial solutions for everyone.

Writing Skills

It's crucial to be able to communicate clearly and concisely in emails as well as more formal documents like reports or proposals.

I learned this the hard way, receiving "TL;DR" (Too Long; Didn't Read) as a response to my long reports. So, I changed my long reports to a five-sentence summary, outlining the most important points and, according to my past self, leaving aside contexts and nuances that I considered important and fundamental.

Today, after all those TL;DRs, I consider an email longer than a paragraph or five or six key points to be utterly unnecessary, a waste of time for both the writer and the recipients.

To improve in this skill, you can follow some of these tips:

- Use clear and concise language, be direct, and avoid using obscure or confusing words.

- Use an appropriate tone, adapting the message to the audience and its purpose.

- Take the necessary time to review and edit the message to ensure it's free of errors and expressed in a clear and concise way.

Verbal Communication Skills

It's essential to be able to communicate effectively in both informal meetings and formal presentations.

The more experience I acquire, the fewer technical terms I use when speaking. This is common for most professionals in this industry, especially if they interact with people in different roles, with different experiences, or simply interlocutors who don't have context for the product language.

Interpersonal Skills

Being able to work effectively with a wide range of people and establishing positive relationships with them is fundamental.

This is closely related to the verbal and written communication I mentioned earlier. It's necessary to understand people's profiles, the context of the stakeholders, and your customers' problems, and to have the necessary emotional intelligence to communicate with all these people in a positive and productive way.

LEADERSHIP AND MANAGEMENT SKILLS

Currently, there's a lot of talk about leadership, and there's little new to add to what has already been said. Still, as I mentioned earlier, you will have to learn to be the leader with each new team. The number of challenges a leader faces is equivalent to the number of people they lead and their connections with each other. In other words, if you lead five people, you'll have a minimum of twenty-five challenges ahead.

You must be capable of leading and motivating your team and managing relationships with other departments to ensure the product's objectives are met. Also, you'll have to make tough decisions, manage the product lifecycle, and make adjustments accordingly.

- Set clear and achievable goals for the team and ensure everyone understands how they can contribute to them.

- Encourage open communication and collaboration so the team can work effectively.

- Provide guidance and support to team members to help them develop and achieve their goals.

- Establish clear responsibilities and roles for each team member to ensure good coordination and avoid conflicts.

- Make effective, data-based decisions to guide the team towards success.

- Manage time and resources effectively to ensure they are used optimally and the team's goals are met.

ANALYSIS AND STRATEGIC THINKING

Strategic thinking is crucial for decision-making and long-term planning. This involves considering how our current actions will affect our future and how we can reach our goals.

Despite the thousands of tools available today for data analysis, such as Google Analytics, Aptitude, Smartlook, Hotjar, Segment, Tableau, and Periscope, which I try to use, in the end, having specialized knowledge in spreadsheets proves to be fundamental for data analysis, management, and even to build a firm opinion on the decisions you need to make.

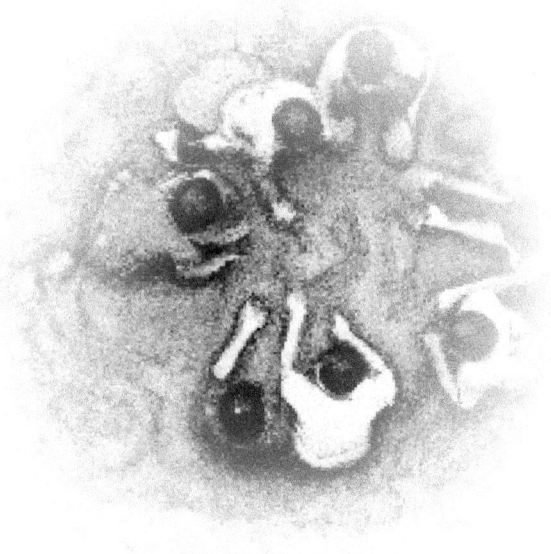

IT'S TIME TO PLAY

We've completed the initiation tutorial in this game of becoming product managers. At this point, we already know what the main skills, methodologies, and responsibilities are that we will have to learn throughout these seven levels.

It's important that, based on everything you've read, you introspect your personality and skills, both technical and soft, and ask your environment to objectively know yourself, so you can focus on developing the weaker side.

I don't want this to be a book full of concepts and definitions, but this introduction, which has taken me longer than I would like, I believe is completely necessary so that when you begin the following levels, you are prepared and familiar with a world that can be extremely complex, but at the same time, will be full of incredible adventures that will give you very powerful, almost addictive, satisfaction, where the only question is why you didn't start this path to becoming a successful product manager earlier.

ASP

`Response.Write("Hello World")`

LEVEL 1
DEFINING THE RIGHT PRODUCT

STRATEGY DEFINITION AND FOCUS OF YOUR PRODUCT

When I was a child, no older than eight or nine years, I was given my first bicycle. I still remember that incredible feeling of thinking it was a spaceship and I could go wherever I wanted, the smell of new rubber from its tires, and the sensation of unparalleled freedom. The first day I set off and from so much back and forth I had marked the asphalt on the whole street; the second and third day the same, and by the end of the week, that incredible feeling had faded into the boredom of not knowing where to go, nor why nor with whom to do it.

I didn't take long to discover that, without a clear destination, having a bicycle was not of much use. I found its true value for me, which was that it would take me to where I needed to be when I needed it, and I went from marking the asphalt of my home street to traveling all over the city, making it indispensable for achieving my objectives.

Thus, having a clear strategy and focus give meaning to the company, the product, and therefore to the PM, who without it, a company is nothing more than a vehicle without a clear direction.

HOW TO DEFINE THE VISION AND PURPOSE OF YOUR PRODUCT?

Let's start with the basics, a purpose is the motive or reason for being of something, while a vision is a mental image of what you want the future of something to be.

A purpose is something deeper and more enduring than a vision, which can be more specific and change over time. For example, the purpose of a company might be to improve people's quality of life, while its vision might be to become the leading company in its sector globally.

Having a clear and coherent vision for your product is fundamental, as this will help guide your decisions and maintain focus. To define the vision and purpose of your product, you could consider questions such as: what is the problem we are trying to solve? What benefits do we offer our customers? What results do we hope to achieve in the long term?

Programming, designing or any kind of activity related to a technical specialization is like building a building: you need a detailed blueprint and a list of necessary materials and tools to carry out the plan. In the same way, defining the vision and purpose of a product is like having a long-term blueprint and a list of objectives and requirements to carry out the project.

When programming or designing a product, you must take into account aspects such as usability, scalability, and security of what you are building. Similarly, when defining the vision and purpose of a product, one must consider aspects such as relevance to the target audience, technical and financial viability, and long-term sustainability.

As a specialist, you must have technical skills to be able to execute the construction project efficiently and according to market quality standards. As a PM, you need technical skills to understand the needs and challenges of a product and make informed decisions.

Being a specialist, whether as a programmer or another activity within the value chain of a tech project, and defining the vision and purpose of a product have many points in common: both require careful planning, a deep understanding of the project's needs and challenges, and technical skills to carry out the work efficiently. As a specialist, you have a solid base of knowledge and skills that will help you become a *product manager*.

EXERCISE 1: « THE FIRST BOSS »

Let's start with the first challenge of this level, which is to try to create the vision and purpose of a company. If we had to define those for products like Netflix, Amazon, IBM or Microsoft, what do you think they could be? And if you had to create your own product's?

I invite you to open a text document or, if you prefer, take pen and paper, and try to define the vision and purpose of the product you choose; it can be one of the examples I gave before and compare them with mine, or you can try with your own product. The objective of this exercise is for you to understand in detail the difference between vision and purpose and understand why it is so important to define them.

On the following page you will find some examples, which will serve as a guide.

Netflix:

- **Vision**: to be the leading entertainment platform worldwide.

- **Purpose**: to provide users with the best content viewing experience, with a wide variety of options and the ability to watch what they want, when they want, and on any device.

Amazon:

- **Vision**: to be the world's largest and most convenient store.

- **Purpose**: to offer users a quick and simple shopping experience, with a wide variety of products at competitive prices and the option to receive their purchases quickly and adequately.

IBM:

- **Vision**: to be the leading technology solutions and consulting company worldwide.

- **Purpose**: to help businesses and organizations solve their most complex problems and challenges through the use of technology and specialized consulting.

Microsoft:

- **Vision**: to be the leading technology and business solutions company worldwide.

- **Purpose**: to help businesses and organizations achieve their goals and increase their productivity through innovative and high-quality technological solutions.

This exercise helps you link what you build as a PM with what your company ultimately wants to achieve. On more than one occasion, I've had to revisit my company's vision and purpose to check if my decisions were pointing in the right direction.

Exercise 1

Exercise 1

Now that we understand what the vision and purpose of our product are for, we need to focus on the problem and the benefits. **What is the problem we are trying to solve? What benefits do we offer to our customers? What results do we hope to achieve in the long term?**

Before building a product or service, it's essential to understand the problem you're looking to solve. For a successful business, you need to understand your customer's pains and how your solution can help alleviate them.

Without understanding the problem, we can't offer a solution. Once we understand it, we can start talking about the benefits of our solution, how it will help improve our customers' lives and how it will help them achieve their goals. In addition to, of course, keeping in mind the expected results to measure the success of our solution, whether it's an increase in sales, higher customer satisfaction, or greater efficiency in internal processes. By understanding the problem, offering benefits, and having the expected results, we can create an effective and sustainable solution for our customers.

EXERCISE 2: « THE SECOND MINI BOSS »

Based on the vision and purpose mentioned earlier, let's look at the second exercise of this level: take the same example you used for the previous exercise and define the problem you solve, the benefit for your customers, and the results you expect to achieve. I took the liberty of writing some examples so you can compare them with yours.

Netflix:

- **Problem:** to offer a content viewing experience that is better and more convenient than traditional television services.

- **Benefits**: a wide variety of content options, the ability to watch what they want, when they want, on any device, and without ads.

- **Expected results**: to become the most widely used entertainment platform worldwide and generate revenue through subscriptions and production of original content.

Amazon:

— **Problem**: to offer a quick and simple shopping experience with a wide variety of competitively priced products.

— **Benefits**: a wide variety of product options, competitive prices, the option to receive purchases quickly and conveniently, and secure payment options.

— **Expected results**: to become the largest and most convenient store in the world and generate revenue through sales and advertising.

IBM:

- **Problem**: to help businesses and organizations solve their most complex problems and challenges through the use of technology and specialized consulting.

- **Benefits**: innovative and high-quality technological solutions and specialized consulting to solve complex problems.

- **Expected results**: to become the leading company in technology solutions and consulting worldwide and generate revenue through sales and consulting services.

Microsoft:

- **Problem**: to help businesses and organizations achieve their goals and increase their efficiency through the use of technology and software solutions.

- **Benefits**: a wide variety of software options and technological solutions for different needs and sectors, ease of use, and integration capabilities with other technologies.

- **Expected results**: to become the leading company in software and technological solutions worldwide and generate revenue through sales and software licenses.

Exercise 2

Exercise 2

HOW TO SET OBJECTIVES AND SUCCESS METRICS FOR YOUR PRODUCT USING THE OKR METHODOLOGY?

There is nothing more important for a company than having defined objectives, both in the long and short term. And not only defined, but also communicated, updated, and constantly measured. The difference between managing objectives efficiently and not doing so can cost you your business.

We have previously defined the vision, purpose, problems to solve, benefits, and expected results because they are the starting point for setting objectives and success metrics and for measuring the progress of your product.

A commonly used methodology for setting objectives and metrics is called OKR (objectives and key results), created by Intel and used by companies such as Google, among others.

This involves setting ambitious and measurable objectives, along with key results that allow you to quantify progress towards those objectives.

OKR AT A GLANCE

— Objectives should be ambitious and feel a bit uncomfortable, that is, it should generate uncertainty about the possibility of fully achieving them.

— Key results should be measurable and easy to rate with a number, which is generally between 0.0 and 1.0.

— OKRs must be public for everyone in the organization, and each department must be aware of the objectives and key results of the other departments.

— A score of 60% or 70% should be considered a positive rating if the OKR is met; if 100% of the objectives are consistently achieved, it is because they are not ambitious enough, and larger objectives should be defined.

— Low scores should be analyzed as data to help refine the next quarter's OKR and not to penalize the team. They are not synonymous with performance evaluations and should not be used for that purpose.

I have used various methodologies for defining and tracking objectives, but none have worked as well across a company as OKRs

Their **main characteristics** are:

Flexibility

They can be applied to any area of the company, from the sales team to the marketing team, and can be adapted to the culture and needs of each organization.

Focus

It helps companies maintain focus on what is most important and achieve more ambitious objectives by setting clear and measurable goals.

Alignment

It allows companies to set objectives aligned with the organization's long-term vision and strategy, which facilitates decision-making and progress tracking.

Transparency

It promotes transparency in the objective-setting process and allows all team members to have a clear view of what is being tried to achieve.

Adaptability

It enables companies to quickly adapt to market changes and adjust their objectives accordingly, giving them a competitive advantage.

A fundamental part, if not the main one, is that both the company and **all its departments must have their OKRs defined at the level of**

responsibility that corresponds to them, redefine them per quarter, and align them with the objectives of each area; to make it simple, achieving 60% of the objectives in each area should automatically complete the overall objective set by the management.

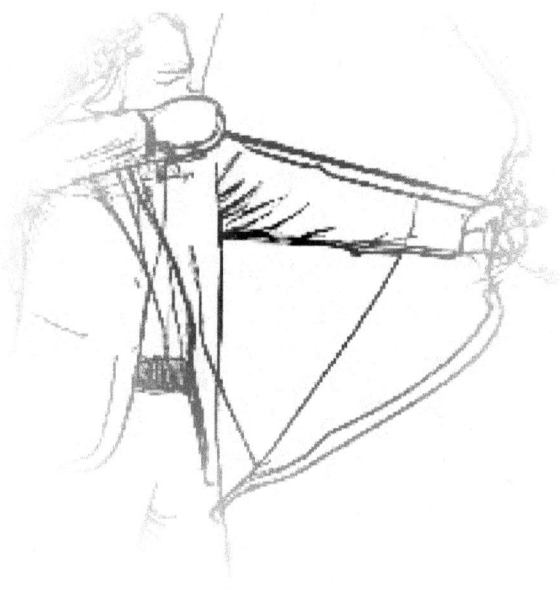

STEPS TO DEFINE OKR CORRECTLY

To correctly define an objective, we must follow a series of steps to ensure that they are correct and feasible, we will follow the following sequence which I will explain in detail later:

Each team leader from each area of the company is asked to draft about ten possible objectives for the management team to evaluate. After an initial filter and analysis of short and long-term strategies, the management team will propose a list of five objectives, two of which we will have to commit to achieving, and the other three will be optional or less important.

Once the overall objectives for the quarter have been defined and agreed upon, the objectives for each department must be defined, and the key results for each objective. Each team leader performs exactly the same action, but one step lower, until all operational teams are reached.

This objective definition methodology gives the entire company the opportunity to participate in the definition of objectives. Of course, all team suggestions may not be heard, and objectives can be defined and consolidated without team feedback, although for obvious reasons, I do not recommend this.

Once the objectives have been defined, we must define the key results for each one, and agree on what the tracking metrics are.

After all this process, the objectives are communicated to all departments, they are reviewed weekly, and the results are shared at the end of the quarter to start again with the same process of defining the new OKRs.

COMMON MISTAKES
WHEN WRITING OUR OKR

Poor communication of OKR

Setting ambitious objectives requires careful communication within teams. If your product depends on the goals of another team, make sure you understand their goal-setting philosophy.

BaU (business as usual)

OKRs are often written based on what the team believes it can achieve without changing anything it is currently doing, rather than what the team or its customers really want. Set aside low-priority efforts and allocate resources to the most important OKRs. There are some objectives that will remain the same from one quarter to another, and this is fine if that objective is always a high priority, but key results should evolve to drive the team to continue innovating and to be more efficient.

Low-value objectives

OKRs should promise clear business value, otherwise, there is no reason to spend resources on them. "Low-value objectives", even if fully achieved, will not be of much importance to the organization. Ask yourself, could the OKR score a 1.0 under reasonable circumstances without providing direct organizational benefit? If so, rewrite the OKR to focus on tangible benefit..

Insufficient key results for each objective

If the key results for a given objective do not represent everything necessary to fully achieve it, unexpected failure may occur, they can generate delays or there may be blocks that were not taken into account in their definition.

To give you the confidence you need to use OKR as an objective management methodology compared to others, these are all the methodologies I have used throughout my career, either out of obligation in some company or out of simple curiosity to see if it was productive:

A very important clarification for the purists on this subject, many times I use the word methodology instead of framework and although the differences are clear, the reason is that I am interested in promoting detailed and specific planning techniques, especially if you are starting in this role, since a framework gives you much more flexibility but brings the risk of straying from what's important easily.

SMART

This is based on setting goals that are specific, measurable, achievable, relevant, and time-bound. SMART, an acronym for these five key characteristics of a well-defined goal: Specific, Measurable, Achievable, Relevant, and Time-Bound. SMART goals help focus attention on what's important and increase the likelihood of achieving positive results.

Hoshin Kanri

This Japanese methodology is used to set long-term objectives and establish action plans to achieve them. Objectives are divided into strategic, tactical, and operational, and metrics are used to measure progress.

It is divided into three main stages: planning, implementation, and monitoring. During the planning stage, the strategic objectives of the company are defined, and objectives at the team and individual level are set. During the implementation stage, the necessary actions to achieve the set objectives are implemented. Finally, during the monitoring stage, the objectives are periodically tracked to ensure their fulfillment.

Balanced Scorecard Metrics

This involves establishing a set of key metrics that measure an organization's performance from four perspectives: financial, customer, internal processes, and learning and growth. These metrics are used to measure progress towards the organization's long-term objectives.

The methodology is based on the idea that a company cannot be adequately evaluated only by its financial results and that a wide range of factors contributing to the organization's long-term success needs to be considered.

North Star Metrics

This involves setting a key metric, also known as a "North Star," that represents the long-term objective of a product or service. This metric is used to calculate progress towards the ultimate goal and can be broken down into more specific metrics for a more detailed view. I find it particularly useful due to its focus on identifying a single key metric that represents the core value of the product to the user.

It has a user-centered approach, seeking to understand what motivates users to interact with the product and how it can be improved to meet their needs and desires. Although I find it very complicated to correctly choose this metric, especially if you don't have experience yet as a PM, it is carefully selected to reflect the core value that your product offers, and is measured and optimized at all times.

It allows product teams to focus on the user experience, rather than focusing on secondary or performance metrics. It helps teams prioritize resource allocation and product decisions more effectively, as they focus on a single metric that really matters.

5W Method

This approach is based on answering the questions "what?", "who?", "when?", "where?" and "why?" to define a goal clearly and concisely. It is a tool used in research, planning, and problem-solving that allows teams to gain a comprehensive and detailed understanding of a situation or problem.

Value Chain Method

This approach is based on identifying key activities that add value to a product or service and setting objectives based on these activities. Each activity a company carries out adds value to its products or services. It seeks to understand which activities are more valuable than.

It is divided into six main categories:

- **Primary Activities:** production, delivery, and customer support.

- **Support Activities:** human resource management, research and development, technology management, and infrastructure.

- **Input Processes:** purchasing management, information management, and resource management.

- **Output Processes:** sales management, marketing management, and customer relationship management.

- **Development Activities:** include research and development of new products and services.

- **Infrastructure:** technology management, information management, and resource management.

Six Sigma

This is based on the continuous improvement of processes and the elimination of errors through the definition of quantitative objectives and the use of metrics and statistical tools to measure progress. It has been successfully used in a wide range of industries to improve efficiency, reduce costs, and increase customer satisfaction.

It is composed of five phases or stages: Define, Measure, Analyze, Improve, and Control (DMAIC). In the first phase, Define, the problem or opportunity for improvement is identified and the objectives and goals to solve it are set. In the second phase, Measure, relevant information is collected and analyzed to understand the problem and its underlying causes. In the third phase, Analyze, the information collected is used to determine the root causes of the problem. In the fourth phase, Improve, solutions are developed and implemented to solve the problem and improve the process. Finally, in the fifth phase, Control, the process is monitored to ensure that improvements are maintained in the long term.

COMPARATIVE TABLE OF COMPLEXITY

Methodology	Complexity	Author	Year
OKR	Medium	John Doerr	1994
SMART	Low	Peter Drucker	1954
Hoshin Kanri	Medium	Yoichi Ueno	1960
Balanced Scorecard	Medium	Robert Kaplan and David Norton	1990
North Star	Low	Aaron Ross and Jason Lemkin	2011
Six Sigma	High	Bill Smith	1986
Scrum	Low	Ken Schwaber and Jeff Sutherland	1993
5W Method	Low	Unknown	Unknown
Value Chain Method	Medium	Michael Porter	1980
4P Method	Low	Jerome McCarthy	1960

EXERCISE 3: «THIRD MINI BOSS »

For the third exercise of this level, let's take the work done so far and, based on the vision, purpose, problem, benefits, and expected outcomes, we can create our OKRs. On the following page, you will find examples that can serve as a guide.

For space reasons, I have only detailed the OKRs of each department in the first example, and I have left only the global OKRs in the rest.

Netflix

- **Global Goal:**

  - **Objective**: Become the leading streaming service globally.

  - **Key Result Indicator (KRI):** Increase the number of annual subscribers by 10% over the next three months.

  - **Result (R):** Launch a marketing campaign in all Latin American countries over the next three months.

- **OKR by department:**

  - **Marketing Department:**

    Objective: Increase brand awareness and campaign reach in Latin America.

    Key Result Indicator (KRI): Increase organic social media reach in Latin America by 20% over the next three months.

    Result (R): Develop a content strategy and launch a social media advertising campaign in Latin America over the next three months.

  - **Product Department:**

    Objective: Improve user experience on the platform.

    Key Result Indicator (KRI): Increase platform usage time by 15% over the next three months.

Result (R): Launch a new version of the mobile app with improved features and updated design over the next 3 months.

- **Finance Department:**

Objective: Increase profitability.

Key Result Indicator (KRI): Increase the net margin by 10% over the next three months.

Result (R): Evaluate and negotiate licensing and distribution agreements over the next three months.

- **Operations Department:**

Objective: Increase content processing capacity.

Key Result Indicator (KRI): Increase content streaming speed by 20% over the next three months

Result (R): Modernize the storage and content transmission infrastructure over the next three months.

IBM

- **Objective**: Transform the information technology industry through innovation.

- **Key Result Indicator (KRI):** Increase the number of approved patents by 15% annually over the next three months.

- **Result (R):** Invest in the development of new artificial intelligence and machine learning technologies over the next three months.

Amazon

- **Objective**: Become the largest and most successful e-commerce company in the world.

- **Key Result Indicator (KRI):** Increase the number of online orders delivered within 24 hours by 20% over the next three months.

- **Result (R):** Expand the network of distribution centers and improve shipping process efficiency over the next three months.

Microsoft

- **Objective:** Be the leader in the business technology market.

- **Key Result Indicator (KRI):** Increase market share in the business software segment by 10% over the next three months.

- **Result (R):** Launch a new product line for project management and collaboration over the next three months.

I hope that with these examples, you can get a concrete idea of what an OKR aims to solve. If an objective is set, it is fundamental to know how we are going to achieve it and how we are going to measure it, and that all teams are informed.

Exercise 3

Exercise 3

HOW TO DEFINE YOUR TARGET AUDIENCE AND UNDERSTAND THEIR NEEDS AND PROBLEMS?

Like any successful company, it is essential to understand who your target audience is and what their needs and problems are. The way you define your audience and understand their needs and problems is key to effectively directing your product or service to people who are willing to buy it.

First, you need to define who your target audience is. This may seem like a simple task, but it is often more difficult than it appears. You need to consider factors such as age, gender, education, income level, and geography, among others. Once you have defined it, you must research to understand their needs and problems.

To understand the needs and problems of your target audience, you must conduct market research. This can include surveys, interviews, and focus groups. Through these tools, you can gain valuable insight into what your target audience is looking for in a product or service and what their biggest challenges and concerns are.

When you have collected and analyzed the information, you should use it to develop a detailed profile of your audience and their needs. This profile will help you better understand your potential buyers and adapt your product or service.

In summary, through market research and data analysis, you can develop a detailed profile of your target audience and use this information to adapt your product or service to meet their needs.

Although it seems like a one-time task when you create your product, ideally, you should constantly update your definition of your audience, as different aspects of our product or service meet different nuances of the buyers, and generally, we have many different audiences per product.

There is a very interesting debate on this topic. In fact, when I have requested feedback from my colleagues on this issue and commented on my point of view, there has been a nearly 50% split over who should define this.

On the one hand, a wide variety of product managers delegate this task almost entirely to the marketing team, especially UX specialists who create the user persona and do the corresponding research, providing the PM with the result of this.

And on the other hand, those who, like me, prefer to be involved in the process and participate in the research and definition of each user persona and each target audience.

The context and understanding that extreme knowledge of your users/customers gives you is critical for decision-making.

WHAT IS A USER PERSONA?

A user persona is a fictional profile of your ideal customer that represents a specific segment of your target audience. It is used to help businesses better understand their customers and develop products and services that best meet their needs. A user persona is based on market research and real data and is used to guide strategic decisions in areas such as user interface design, advertising, and marketing; most of the functionalities or services of your product will be aimed at a specific user persona.

To create a user persona, you need to gather information about your current and potential customers. This can include demographic data, such as age, gender, education, and income level, as well as information about their interests, needs, and challenges.

Here's an example of a detailed profile of your user persona:

- Name: Juan Pérez
- Age: 35 years old
- Gender: masculino
- Education: Degree in Business Administration
- Occupation: Marketing Manager
- Income Level: € 70,000 - € 100,000
- Family: Married with 2 children
- Interests: Sports, Travel, and Technology
- Needs: Need for an effective marketing solution to increase sales
- Challenges: How to reach a wider and more diverse audience

Once you have developed your user persona, you should use it to guide your strategic decisions. For example, when designing a user interface, you should take into account the needs and challenges of your user persona to ensure that your product or service is easy to use and provides an excellent user experience.

You can also use your user persona to guide your marketing and advertising efforts, to ensure that you are reaching the right people with the timely message.

As I mentioned before, there are countless techniques to conduct market analysis, some simpler and others more complex or expensive.

Some of the most common ones are:

Surveys:

This involves sending structured questions to a representative group of people to get information about their opinions, behaviors, and preferences. Surveys can be conducted in person, over the phone, or online.

The process of conducting surveys is a fundamental part of defining the target audience of a product, they are an effective tool to gather valuable information about the desires and needs of consumers.

The first stage in the survey process is to define the questions that need to be asked, these questions should be clear, precise, and focused on the investigation of the target audience. For example, if you are developing a product for seniors, it is important to ask about their buying habits, brand preferences, and specific needs.

Once the questions have been defined, you must select the appropriate survey method, online surveys are a popular option due to their accessibility and speed, while in-person surveys allow for more personal interaction and a higher response rate. Choosing the method that provides the most accurate and reliable information for the type of user you need to survey is essential.

With the questions defined, you should think about how you will collect the data, the sample must be representative of the target audience, and the response rate must be high enough to get an accurate picture of the market.

Focus Groups:

They are based on bringing together a small group of people to discuss a particular topic or product. Focus groups are moderated by an expert and can be recorded for later analysis.

The first stage in the focus group process is selecting the right participants. This includes ensuring that participants represent the target audience of the product and have an interest in the topic being discussed. For example, if you are developing a product for children, it's important to include parents and children in the focus group.

Once the participants are selected, it's important to define a script for the discussion that includes guiding questions and a clear structure to keep the focus on the research objectives. During the discussion, it's important that the facilitator maintains an open and neutral environment so that participants feel comfortable sharing their opinions and perceptions.

After the discussion, it's important to analyze the collected data and use it to improve the product. This may include identifying product strengths and weaknesses, pinpointing areas for improvement, and adjusting the marketing strategy.

Individual Interviews:

These involve having a deep conversation with a person about a particular topic or product. Individual interviews can be structured, semi-structured, or open-ended.

It is the preferred method for more than 80% of the product managers who have been interviewed for this book. A 15-minute conversation where trust is the central point allows them to not only understand in depth the problems that the product seeks to solve for this particular segment, but it's also a very common consultative sales format. *Business progresses at the speed of trust.*

Observation:

This involves observing people's behavior in their natural environment to understand how they interact with a product or service. Observation can be natural or controlled.

There are different ways to conduct an observation, such as participant observation, where the observer is an active part of the group being observed, or non-participant observation, where the observer is a passive spectator. The choice of methodology will depend on the objectives of the research and the context.

It's important to note that observation is non-invasive and is carried out discreetly so as not to influence consumer behavior. In addition, it's important to have a clear observation guide that includes the objectives of the research, questions to answer, and a list of behaviors and actions to observe.

If you have doubts about when to use one technique or another, I present you with a comparative table of the market research techniques mentioned:

Technique	Advantages	Disadvantages
Survey	Allows obtaining direct and specific information from a certain population. It's quick and relatively inexpensive.	Depends on the quality and reliability of the selected sample. There may be biases or lack of honesty in the responses.
Interview	Allows obtaining detailed and deep information about a specific topic. It helps to establish relationships and understand the context of responses.	Requires time and effort to prepare and conduct the interviews. There may be biases or lack of honesty in the responses.
Focus group	Allows obtaining a more complex and diverse perspective on a topic. It helps to generate innovative ideas and solutions.	Requires time and effort to prepare and conduct the focus group. There may be biases or lack of objectivity in discussions.
Observation	Allows obtaining close and real information about consumer behavior. It doesn't require the intervention of the researcher, which can lead to more authentic observations.	Can be costly and require a great deal of time. It doesn't allow controlling variables. It can be difficult to analyze the information obtained.

EXERCISE 4: « CHALLENGING THE FIRST LEVEL »

Following the previous exercises, our fourth and last exercise for this level is to describe the target audience and at least two user personas, taking into account everything we know so far; I leave some examples as a guide on the following page.

Netflix

Our target audience are online entertainment consumers worldwide who are looking for a broad selection of series, movies, and documentaries, through an easy-to-use and accessible platform.

User persona:

- Maria, a 22-year-old college student who seeks a variety of series and movies to watch while studying or relaxing in her spare time.

- Jose, a 35-year-old family man who seeks family-friendly shows and movies to watch with his wife and children on weekend nights.

- Alex, a 60-year-old retiree who seeks educational documentaries and travel shows to watch in his spare time

IBM

Our target audience are companies of all sizes and sectors looking for advanced technological solutions to improve their efficiency and productivity.

User persona:

- Carmen, IT director of a medium-sized e-commerce company, seeks technological solutions to improve efficiency in the order process and business data analysis.

- Pedro, Human Resources manager at a large construction company, seeks a technological platform to improve employee management and productivity.

Amazon

Our target audience are online consumers worldwide who are looking for a wide selection of products at affordable prices and with fast delivery.

User persona:

— Marta, a 30-year-old housewife, seeks a wide variety of products for her home at affordable prices and with fast delivery.

— Luis, a 45-year-old frequent business traveler, seeks a wide variety of travel and electronic products at affordable prices and with fast delivery.

Microsoft

Our target audience are businesses and consumers worldwide seeking high-quality and advanced technological solutions for their computing and business needs.

User persona:

- Ana, the CEO of a small software development company that seeks high-quality solutions for her software development and business needs.

- Carlos, a 25-year-old computer science student who seeks advanced computing solutions for his projects and school assignments.

Exercise 4

Ruby

`puts "Hello World"`

LEVEL 2
BUILDING THE RIGHT PRODUCT FAST AND CHEAPLY

FEEDBACK:
ENGAGING CUSTOMERS

To illustrate the process of engaging customers and obtaining their feedback, one could compare it to the process of programming. Imagine, just as a programmer writes code and then tests and debugs their application to ensure it operates correctly, a product manager must engage customers and collect their feedback to improve the product.

In the same way a programmer uses tools like debuggers and unit tests to identify and correct errors in their code, the product manager can utilize market research techniques and metrics to understand their customers and adjust the product accordingly.

For instance, a programmer might write a function that calculates the average of a set of numbers. If during testing they realize the function doesn't handle negative numbers correctly, they can modify their code to fix the error. Similarly, a product manager might launch an initial version of a product and then use techniques like customer surveys and interviews to gather feedback and determine how to enhance the product in future iterations.

FEEDBACK IN VARIOUS BUSINESS STAGES

Feedback is a key process in product development and business growth. At each stage of a product lifecycle, feedback is crucial to make informed decisions and ensure the product aligns with market needs.

We can define three key stages in a business's life: the early stage, where the business is born and defined; the continuous iteration stage, which may include significant changes in segments, functionalities, and services, and is generally the phase where a product either dies or fully reasserts itself; and finally, the scaling or globalization stage, where we take the product or service towards a more stable or corporate state.

During the product creation and definition stage

Feedback is obtained through market research and validating product ideas with potential customers. It is essential to obtain it early and often to ensure that a real and market-desired problem-solving product is being created.

During the continuous improvement stage

It is obtained through monitoring and analyzing product performance metrics, as well as from existing customers. Implementing a robust system to collect and analyze this feedback is key for the business, as this will allow the product development team to understand what improvements to make and prioritize features.

During the scaling to new markets stage

Understanding the differences between markets and adapting the product and marketing strategy accordingly is essential. Gathering information from both existing and potential customers in these new markets, and using this feedback to adjust value proposition, pricing, and marketing is crucial to ensure success in these new markets.

Feedback is a critical element at every stage of the product and business lifecycle. Without it, the product development team may make decisions based on incorrect assumptions about the market and customers. Conversely, with a strong feedback culture, a company can develop highly relevant products and successfully scale in emerging markets.

A few years ago, I had the challenge of participating in the scaling process of a startup in Latin America, taking a SaaS leader in Argentina, Chile, and Mexico to the Brazilian market. The first mistake we made was thinking that this market would behave like its South American neighbors. We didn't conduct any surveys, user personas, listen to feedback from potential customers, and overlooked all internal sales team opinions that would handle that market. Once the product was launched, 90% of the leads dropped at the payment stage. We were losing all the investment in this new market and didn't know why.

We only had to talk to two companies to realize that the primary reason was the payment options we offered. We didn't take into account regionalizing the payment process, we weren't offering the region's most used payment method, leading to consumer abandonment of the process and losing all invested in marketing and sales to capture them.

It took us no more than two weeks to regionalize the entire payment process, and a year later, the company was acquired by a Brazilian corporation.

The value of talking to your customers regularly, systematically, and sustainably is immeasurable. So much so that it should be rooted in every company's culture.

HOW TO INVOLVE CUSTOMERS IN THE PRODUCT DESIGN AND DEVELOPMENT PROCESS?

Customers are the primary source of feedback to understand their needs and problems, and their involvement in the decision-making process can be key to a product's success, as I have already mentioned. Moreover, involving customers in the product design and development process helps foster customer loyalty and satisfaction, which can positively impact the product's long-term success.

You can simply pick up the phone and call your customers or potential customers and have open conversations, but then it will be very complex to know the good and the bad that conversation has left you, so we must use techniques that help us systematize this process:

Establish effective communication channels with customers

Defining the effective channel to obtain customer feedback depends on several factors, including the objective of the feedback, the size and composition of the target audience, the budget and resources available, and the time available for feedback.

It's crucial to be proactive in seeking customer feedback and be willing to listen to and consider their opinions and suggestions.

Offer incentives to customers to participate in the product design and development process

This involves offering rewards, discounts, or the chance to be a beta tester or a design team member, among thousands of other options.

However, it's important to note that incentives should be fair and equitable to customers, and they should not be perceived as manipulative. You must ensure that incentives do not negatively affect the feedback, such as making customers provide false or untruthful feedback to obtain an incentive.

Product managers should carefully consider incentives when choosing a channel to obtain customer feedback and ensure that their use improves the product's quality and experience.

Use market research techniques and metrics to better understand customers

The combination of market research along with metric analysis can provide a complete picture of customers and their needs, allowing product managers to make informed decisions about how to enhance and develop their products. Furthermore, these techniques and metrics can be continuously monitored to adapt to changes in customers' needs and preferences as they evolve.

Consider using focus groups or individual interviews to gather qualitative feedback

I have previously mentioned focus groups and individual interviews as a fundamental tool to define the target audience, but they are extremely useful to obtain ongoing feedback about your product.

These methods allow for a deep dive into customers' thoughts and opinions and can provide valuable insights that cannot be obtained through surveys or quantitative metrics.

Create an easy and convenient system for customers to provide feedback

Online forms, a customer service phone line, online chat, etc.

One practice that has helped me a lot is that the entire product team has access to customer opinions and knows how to apply them to improve the product. To each proposal I make, I always add no fewer than three customer opinions that validate the functionality, product, or service to reinforce the reasons for implementing it. I do the same with my teams, every time someone proposes a new functionality, I ask them to bring me user feedback where they have, in some way or another, requested what they propose.

DIFFERENCES BETWEEN DOING B2B AND B2C PRODUCT

It's important to keep these differences in mind when approaching the product management process and adapt the approach to the specific characteristics of each type of business.

In the case of B2C (business-to-consumer) businesses, it is common to use surveys and online feedback to collect customer feedback. These techniques are accessible and allow obtaining a large amount of data in a short period of time. Additionally, being online, they allow feedback from a wide and diverse audience. However, it's important to bear in mind that online feedback may be less accurate and biased, as it only includes those customers who decide to participate.

On the other hand, in the case of B2B (business-to-business) businesses, it's common to use in-depth interviews and focus groups to gather customer feedback. These techniques allow obtaining more accurate and detailed feedback, as they allow delving into customers' needs and preferences. Furthermore, as interviews and focus groups are in person, they enable a closer relationship with customers and more reliable feedback. However, these techniques can be more costly and require more time to be implemented.

The main differences between doing **B2B and B2C** products are:

Target market

In the B2B approach, the product is sold to other companies; whereas in the B2C approach, the product is sold to end consumers. This means that the target market is different in each type and requires different marketing and communication techniques.

Purchase decision process

In the B2B approach, the purchase decision process tends to be more complex and require the participation of several people in the buying company. In B2C, the purchase decision process is usually simpler, involves fewer people, and takes less time.

Type of product

In B2B, there are usually more technical and specialized products than in B2C. This means that the product team needs to have a deeper understanding of the product and its application in the market.

Product life cycle

In the B2B approach, the product life cycle tends to be longer and require a higher commitment from the customer. In B2C, the product life cycle is usually shorter and requires less commitment from the customer.

Customer relationship

In B2B, the relationship with the customer tends to be longer-lasting and more personalized due to the customer's higher commitment and the size of the orders. In the B2C approach, the relationship with the customer is usually more transactional and less personalized.

In the B2B approach, the product management process may require a greater focus on research and analysis due to the complexity of the market and the purchase decision process. There may also be a greater emphasis on the customer relationship and product customization, and from a marketing strategy perspective, a more technical and specialized approach may be needed due to the nature of the product and the more complex purchase decision process. There is generally a greater emphasis on the customer relationship and customization of the marketing message.

In the B2C approach, the product management process may require a greater focus on creativity and innovation due to the shorter product life cycle and the more competitive market. And generally, there is a greater emphasis on promotion and communication with the end consumer. The marketing strategy may need a more creative and engaging approach due to the nature of the product and the more competitive market, and a greater emphasis on promotion and communication with the end consumer is usually crucial.

EXERCISE 5: « THE UNEXPECTED ENEMY »

Following the previous exercises, this time we have to create a survey that can help you better understand your current or potential customers, taking into account the two mentioned approaches, B2C and B2B. On the next page, I leave some examples that can serve as a guide.

Netflix

1. How long have you been a Netflix user?

2. What are your favorite programs or movies on Netflix?

3. How would you rate the quality of image and sound on Netflix?

4. How would you rate the content selection on Netflix?

5. Have you experienced any technical problems using Netflix? How was it resolved?

6. What kind of content would you like to see on Netflix in the future?

7. Are you interested in receiving personalized content recommendations on Netflix?

8. Are you interested in using additional Netflix features, such as the ability to download content to watch offline?

Amazon

1. How long have you been a Netflix user?

2. What are your favorite programs or movies on Netflix?

3. How would you rate the quality of image and sound on Netflix?

4. How would you rate the content selection on Netflix?

5. Have you experienced any technical problems using Netflix? How was it resolved?

6. What kind of content would you like to see on Netflix in the future?

7. Are you interested in receiving personalized content recommendations on Netflix?

8. Are you interested in using additional Netflix features, such as the ability to download content to watch offline?

Microsoft

1. How long have you been a Netflix user?

2. What are your favorite programs or movies on Netflix?

3. How would you rate the quality of image and sound on Netflix?

4. How would you rate the content selection on Netflix?

5. Have you experienced any technical problems using Netflix? How was it resolved?

6. What kind of content would you like to see on Netflix in the future?

7. Are you interested in receiving personalized content recommendations on Netflix?

8. Are you interested in using additional Netflix features, such as the ability to download content to watch offline?

IBM

1. How long have you been a Netflix user?

2. What are your favorite programs or movies on Netflix?

3. How would you rate the quality of image and sound on Netflix?

4. How would you rate the content selection on Netflix?

5. Have you experienced any technical problems using Netflix? How was it resolved?

6. What kind of content would you like to see on Netflix in the future?

7. Are you interested in receiving personalized content recommendations on Netflix?

8. Are you interested in using additional Netflix features, such as the ability to download content to watch offline?

Exercise 5

Exercise 5

Go

```
fmt.Println("Hello World")
```

LEVEL 3

BUILDING THE RIGHT STRUCTURE

TURNING FUNCTIONAL AREAS INTO MULTIFUNCTIONAL ONES

The success of a product largely depends on the synergy and collaboration between two or more key teams. Setting clear objectives and realistic deadlines, providing the development team with all the necessary information so they can complete the work effectively, and working with the marketing and finance teams to promote and sell the product effectively and comply with all regulations that may prevent us from scaling are the fundamental pillars of a product manager's work.

There are many types of structures, and one of the most common is functional, that is, to have staff divided into areas such as: technology, marketing, finance, purchasing, sales, operations, etc. Although I do not recommend this structure because it is the usual one in most companies, it serves us as an example to understand how to relate and collaborate transversally with them to achieve the product objectives.

In these types of structures, each functional area is usually led by an area director, a chapter, or a specialist leader, who will have their objectives and key results and will have to organize their priorities, just as you will have to do with your team.

This is why it is so important to have globally defined OKRs in the company and for each area to have their objectives aligned with each other, in this way it will be very easy for you to collaborate, since, if the OKRs are well defined and correctly communicated, the fact that an area meets theirs should "partly" help you meet yours.

In my experience, the least risky way to work between areas is to force the links that are part of each team to collaborate with each other to produce the expected results, making the development team able to communicate freely and without intermediaries with each of the company's teams; your role will be to collaborate for this to happen, mediating to negotiate, facilitate or help unlock tasks when necessary.

As I mentioned before, the functional structure "works", although it is not ideal. A structure by cells, tribes, squads or multidisciplinary teams is nowadays the one that has given the most results to companies, creating multifaceted, autonomous and totally independent teams from each other, so that there is a specialist for each need that needs to be covered, based on the objectives set for this team; that is, this type of teams seeks to have a wide rotation of its members according to their objectives.

For example, if we aim to have a 2% increase in the conversion rate, we need to create a team of specialists in all areas where we need operational and decision-making muscle.

This type of structure allows us to free ourselves from the pressure of having objectives for each of our areas, so that the whole company adjusts to the objectives it needs to meet each quarter and not the other way around, where many times there are areas without clear objectives and others overloaded and without resources.

PRODUCT DEVELOPMENT TEAMS

Although as a PM you will have a cross-functional relationship with the entire company, the relationship will be much closer with the product development team, which you will have to lead directly:

Set clear and specific objectives for the project

It is important that both teams have a clear and precise understanding of what is expected from the project and how it contributes to long-term business objectives.

Set realistic and reasonable timelines and budgets

It is crucial that both teams agree on the timelines and budgets assigned to the project and that they are realistic and reasonable. It is better to set slightly wider and more conservative timelines and budgets, rather than tight and ambitious ones, as this can lead to team overload and unwanted delays.

Provide the development team with all the necessary information

Ensure you provide the development team with all the necessary information so they can complete the work effectively. This includes technical details and product features, as well as any other relevant information that may be necessary to complete the project efficiently.

Maintain open and regular communication

Maintaining open and regular communication with the development team to ensure you are progressing adequately and to solve any problem or difficulty that may arise will be part of your daily routine, and you must ensure that there is a constant flow of information. This may include regular follow-up meetings, progress reports, and any other form of communication that is appropriate for the project.

Be flexible and willing to make adjustments

Most of the time, unforeseen events or changes in the project will arise that can affect the timelines and budgets. It is important to be flexible and willing to make adjustments in the process if necessary, as long as it does not compromise the quality of the final product.

RELATIONSHIP WITH MARKETING AND SALES TEAMS

Depending on the type of team you have to lead, working with the marketing team or leading a marketing and sales team can be very common for a PM, in a very similar way to coordinating the tasks of a development team, you need to ensure actions are coordinated among the teams:

Establish a clear and solid marketing strategy

It's important to have a clear and solid marketing strategy to effectively promote and sell the product. This strategy should include a detailed plan on how the product will be promoted and sold, who the target customers are, and how they will be reached. If you are not responsible for creating this strategy, you must ensure that one exists and that it has been communicated between the areas.

Work with the marketing team to develop a launch campaign

The launch of a new product is a key opportunity to attract potential customers and effectively promote the product. Work with the marketing team to develop a solid launch campaign that includes a combination of marketing tactics, such as online advertising, social media, events, and promotions.

This is where the PM, in coordination with the development and marketing teams, needs to be creative and unify their potential to create PLG, Growth Hacking strategies, etc., where both teams will need to actively intervene.

Provide the marketing team with all the necessary information about the product

It's crucial that the marketing team has a complete understanding of the product and its features and benefits. Provide the team with the necessary information so that they can effectively promote the product.

A practice that I found very effective is to do small demos of all the functionalities, modifications, and corrections made to the product at the end of each sprint and communicate them to all the teams in the company. In this way, the marketing team always found an opportunity to promote one of them that was not taken into account at the time of its definition and development.

EXERCISE 6: « AN OLD FRIEND »

In accordance with the structures that I have mentioned as recommended for creating product teams, and based on your OKRs, create a multidisciplinary team defining what roles should be included to meet the set objectives. On the following page, I leave some examples that can serve as a guide.

Examples of decentralized and autonomous structures

Team leader: Product manager

Team 1:
> *Product owner*
> *Scrum Master*
> *Technical lead*
> *Backend Developer*
> *Frontend Developer*
> *UX*
> *Marketing Analyst*
> *Designer*

Team 2:
> *Product owner*
> *Customer success*
> *Data analyst*
> *Channel specialist*

Team 3:
> *Product owner*
> *Copywriter*
> *SEO specialist*
> *Designer*

Exercise 6

Exercise 6

PRODUCT LIFECYCLE

In the world of product management, a product's life cycle refers to the different stages it goes through from conception to eventual discontinuation. It can be divided into four main stages:

Idea

This is the first stage where you have an idea or a need that you believe customers would be interested in buying. In this stage, you do market research to understand the problem and validate the idea, identify the target market, and seek a solution that meets the customer's needs.

Launch

In this phase, an MVP (minimum viable product) is built with the aim of launching it to the market to obtain feedback from it. It is important to carry out a proper launch campaign to generate awareness and attract the first customers.

Growth

If the idea is validated and the MVP is successful, then you enter the growth stage. In this stage, the product becomes a profitable product and seeks to expand into other markets. You try to scale the product and look for ways to improve it to better meet customer needs.

Decline

All products have a life cycle, and eventually the product will reach its decline. It is important in this stage to detect signs of decline so that measures can be taken, such as renewing the product or discontinuing it.

In summary, the product life cycle is a continuous process of learning and improvement, centered on the customer and requiring a strategy to carry it out.

A GENERAL APPROACH FOR ALL STAGES

As you may have noticed, the product life cycle is closely related to the Lean concept, and we can use an agile approach to product development that is based on continuous iteration and learning as a product advances or evolves, regardless of the stage we are in.

Instead of following a strict plan, the agile approach focuses on flexibility and the ability to adapt to changes as you advance in development:

Planning

This is the initial stage where the objectives and goals of the project are established. The market is researched and the target audience is identified in order to define the needs and requirements of the product.

Design and Construction

An initial version of the product known as an MVP (minimum viable product) is built according to the previously established requirements, seeking a balance between functionality and development speed. Remember that an MVP does not necessarily refer to the launch of a new product, we can iterate parts of our product with new concepts that are very complex and that to test if they work we need to implement the same iterative concept, fragmenting the product into several MVPs.

Testing

In this stage, tests are done with the MVP with the aim of collecting feedback from customers or users. This is used to improve the product and to plan for the following versions. These tests are commonly called "A/B tests" or "experiments", which turn out to be small modifications to the product to corroborate possible hypotheses.

Iterations

The process does not end with the launch, but is repeated in a series of iterations, in which feedback is used to improve and scale the product.

PROMOTING A CULTURE OF INNOVATION

An innovation culture is an environment in which employees feel motivated and empowered to seek new solutions and opportunities, and where innovative ideas are recognized and supported.

Innovation can come from anywhere within the company, and it is important that leaders establish an environment that allows and encourages creativity and out-of-the-box thinking. This includes promoting a learning mindset, creating a collaborative environment, valuing and rewarding novelty, and fostering a diversity of perspectives and approaches.

Fostering a culture of innovation not only helps improve internal processes and increase efficiency, but can also lead to the creation of new products and services that can drive growth and expansion of the company. In this sense, fostering a culture of innovation can be a key factor in achieving a prosperous and sustainable future for the company.

Establishing an innovation culture

To manage change and effectively deal with failure, it is important to establish a culture within the company that encourages innovation and calculated risk. This can include creating a dedicated innovation team, promoting a "fail fast, learn fast" mindset, and creating a work environment that fosters creativity and collaboration.

Establishing an innovation process

It is important to establish a systematic and structured innovation process. This process may include identifying innovation opportunities, generating ideas, evaluating them, and implementing the selected ideas. Establishing an innovation process can help ensure that time and resources are invested strategically and minimize the risk of failure.

Establishing a contingency plan

It is essential to establish a contingency plan that outlines how unexpected problems or project failure are handled. This includes identifying key inflection points during the project, creating an action plan for handling unexpected problems, and defining criteria for determining when it is necessary to abandon the project. Establishing a contingency plan can help minimize the impact of failure and ensures the team is prepared to handle unexpected problems.

Failure is a natural part of the innovation process and should not be viewed as a sign of weakness. **Instead, it is important to focus on learning from failures and using those learnings to improve in the future**. By effectively managing change and dealing with failure, you can maximize your opportunities for success and ensure the long-term triumph of your innovation project.

DOING PRODUCT IN TRADITIONALIST ORGANIZATIONS

A traditionalist organization is characterized by having a more conservative and traditional approach to decision-making and daily operations. This can significantly affect the product management process and present specific challenges for the product team. This can mean a solid authority hierarchy, a focus on compliance with established norms and procedures, and a tendency to resist change and innovations.

These challenges include:

Difficulties in implementing changes and new ideas

It is more difficult to implement changes and new ideas due to resistance to change and innovation.

Increased time and effort required to make decisions

Decision-making is more complex due to the solid authority hierarchy and the need to follow established norms and procedures. This leads to a slower decision-making process and requires more effort on the part of the product team to convince company leaders of the viability of certain ideas or changes.

Greater resistance to taking risks and trying new things

Due to the more conservative approach of traditionalist organizations, there may be greater resistance to taking risks and trying new things, which can hinder innovation and the development of new products.

Greater emphasis on stability and continuity

It is common for these types of organizations to place greater emphasis on stability and continuity, rather than innovation and change. This can limit the options of the product team and hinder adaptation to changes in the market.

The product management process in a traditionalist organization presents specific challenges due to resistance to change and innovation, additional time and effort needed to make decisions, greater resistance to taking risks and trying new things, and greater emphasis on stability and continuity.

ROADMAPS AND PRIORITIZATION METHODOLOGIES

To begin, let's do a brief review of the most common task prioritization methodologies:

Roadmaps are diagrams that are used to represent the structure and flow of a project. They show the different stages and how they relate to each other.

The Gantt chart is a type of bar chart that is used to plan and control projects. It shows the duration of each task and how it is distributed over time.

The waterfall development model, also known as the waterfall, is a linear model in which a series of sequential stages are followed to complete a project. Each stage is completed before moving on to the next.

All of these tools have their own advantages and disadvantages and should be used according to the context and specific needs of each project. For example, roadmaps are useful for representing the flow of a project, while Gantt charts are useful for planning and controlling the duration of tasks. The waterfall model is suitable for projects with very defined requirements and little change, while other methodologies, such as Scrum, are more suitable for projects with constantly changing requirements.

Let's see a table to identify their advantages and disadvantages:

Feature	Roadmap	Gantt	Waterfall
What is it?	A visual planning technique that uses diagrams to represent tasks and dependencies between them	A planning tool that shows the progress of projects over time.	A sequential project development process in which each phase is completed before moving on to the next.
When to use?	Roadmap is ideal for projects with many interdependent tasks and can be useful at the beginning of a project to plan and organize tasks.	Gantt is useful for monitoring the progress of a project and can be used throughout its entire life cycle.	Waterfall is suitable for projects with well-defined requirements and can be used when exactly what needs to be done and in what order is known.
Advantages	Allows a clear and visual view of the project and how tasks are related to each other.	Shows the project's progress clearly and allows the plan to be adjusted as needed.	t's easy to follow and allows teams to work in a sequential and focused manner.
Disadvantages	It can be difficult to accurately predict when long-term objectives will be completed.	It is difficult to make changes as the project progresses, as tasks are interconnected.	It's hard to make changes as the project progresses because each project phase depends on the previous one.

Generally, when negotiating with stakeholders, they will always demand deadlines for both "deliverables" and results. Although this is changing thanks to the new generations of business management, it is still common for both innovative companies and start-ups, as well as traditionalist ones, to ask you for specific delivery dates.

A practice that is compatible with the needs of both worlds, the more traditional of the stakeholders and the more innovative of product development teams, is to have roadmaps with the expected results of the main objectives to be addressed, and not a defined roadmap of functionality deliverables.

EXERCISE 7: « A SMALL CHALLENGE BEFORE MOVING FORWARD »

With the vision, purpose, benefits, problems, results, your OKR and your teams defined, it's time to plan these objectives in a roadmap of actions and expected results that your teams can carry forward. Keep in mind that a roadmap must make sense, have dependencies, and a start and end for each of the proposed actions and tasks. On the next page, you will find a very basic example that can serve as a guide.

Objective 1

Increase the number of subscriptions in Latin America by 15% during the first semester of 2024.

- **Action 1:** launch marketing campaigns on social networks in Argentina, Brazil, and Mexico (deadline: end of March 2024).

- **Action 2:** offer a free month to new subscribers in Latin America (deadline: beginning of April 2024).

- **Action 3:** add exclusive Spanish and Portuguese content for Latin America (deadline: end of May 2024).

Objective 2

Improve the user experience on the platform by 20% during the second semester of 2024..

- **Action 1:** implement a new, more intuitive user interface (deadline: beginning of July 2024).

- **Action 2:** add the option to download content to view it offline (deadline: beginning of August 2024).

- **Action 3:** improve streaming quality on mobile devices (deadline: end of September 2024).

Remember that the purpose of a result-oriented roadmap is to have a long-term vision and establish specific and measurable objectives to achieve that vision.

We talked about the OKR methodology before and now about roadmaps. To unify both, it is necessary to focus the objectives and key results (OKR) on the roadmap, this implies identifying the actions and tasks necessary to achieve each objective and key result, and adding them to the roadmap.

A common approach is to create a list of objectives and key results at the company level, and then break them down into objectives and key results at the team and product level, which should have a series of actions and tasks associated in the roadmap.

Keep in mind that the OKR methodology focuses on defining long-term objectives and results and measuring progress towards them, while the roadmap focuses on the planning and execution of short and medium-term tasks. Therefore, you should balance both methodologies and make sure that the roadmap is in line with the company's objectives and key results.

If we use one of Amazon's OKR to see what an OKR definition looks like, its corresponding result-oriented roadmap would be something like this:

OKR n.º 1

Increase customer satisfaction by 20% during the next quarter.

Actions:

— Improve online shopping experience. Deadline: end of month 1.

— Offer a better selection of products. Deadline: end of month 2.

— Improve delivery speed. Deadline: end of month 3.

— Offer better customer service. Deadline: end of month 4.

OKR n.º 2

Increase sales by 30% during the next quarter.

Actions:

- Launch social media marketing campaigns. Deadline: end of month 1.

- Offer promotions and discounts. Deadline: end of month 2.

- Expand presence in the international market. Deadline: end of month 3.

- Improve product visibility on the platform. Deadline: end of month 4.

This is just an example of how Amazon could present its OKR and its roadmap to achieve them. The key is to have clear and measurable objectives, and a detailed action plan to achieve them. With the OKR methodology and a well-structured roadmap, Amazon could ensure that it is on the right track to achieve its business goals.

Exercise 7

Exercise 7

Ada

```
Put_Line("Hello World");
```

LEVEL 4

BUILDING THE RIGHT LEARNING

MEASURE, ANALYZE, DECIDE

Lately, there's a lot of talk about data-driven, which summarizes the concept of creating a team culture that makes decisions based on data. We all believe we do it until we analyze our decisions in detail, and we discover a myriad of micro-decisions based on assumptions, both our own and those of others; opinions we form based on our previous experience, which we base on what was and not on what is, and it's that ego that makes us move forward without conclusively checking our options.

"Ego destroys as much as laziness," a slightly cheesy phrase, but one that has become a truth that I can reaffirm every day. We don't always make decisions without data due to ego, but also due to the laziness we feel when conducting the necessary research when it comes to small decisions, and this is due to not having a culture and habit of measuring and analyzing our hypotheses, but everything starts when we ask ourselves: can I substantiate what I'm saying with data?

Before delving further into this topic, I believe it's vital to mention five books that you must read if your goal is to lead a product team:

Web Analytics 2.0, by Avinash Kaushik, provides a detailed guide on how to measure a website's performance and use that data to improve user experience and increase conversions.

Measuring the Networked Nonprofit, by Beth Kanter and Katie Paine, focuses on how nonprofit organizations can measure the impact of their social media efforts and use that data to improve their strategy.

Data-Driven: Creating a Data Culture, by Hilary Mason and DJ Patil, provides a guide for creating a data culture in a company and using that data to make informed decisions.

Digital Analytics Fundamentals, by Google Analytics Academy, is a free book that provides an introduction to the techniques and tools used to measure a website's performance.

Effective Data Storytelling, by Anil Maheshwari, focuses on how to tell stories through data to communicate information clearly and persuasively.

We can divide metrics into two main groups: qualitative and quantitative. To understand these two groups of metrics, we can say that quantitative ones are those that can be measured and expressed in numbers, while qualitative ones are those that cannot be measured numerically but are based on the quality or characteristic of something.

Quantitative metrics are useful for measuring progress and performance over time, as they allow comparisons and set numeric objectives. Some examples include the number of sales, the number of website visits, and the number of app downloads.

Some of these metrics are:

Conversion rate (CR)

It's the percentage of visitors who perform a specific action on the platform, such as completing a contact form or making a purchase. For example, on Amazon, the conversion rate would be the percentage of people who make a purchase on the site.

CR = (number of conversions / number of visits) × 100

Where "conversions" refers to the number of times a user performs a specific action on your website (such as making a purchase, completing a form, etc.) and "visits" refers to the total number of visits to the website. The result is expressed as a percentage.

Cost per acquisition (CPA)

It's the cost the company incurs to acquire a new customer. For example, at Netflix, the CPA could be the money spent on advertising to attract new subscribers.

CPA = total cost of the campaign / number of conversions

Where "total cost of the campaign" refers to the money spent on the advertising or marketing campaign (including ads, landing page customization, etc.), and "number of conversions" refers to the number of times a user performs a specific action on your website (make a purchase, fill out a form, etc.). The result is expressed in currency.

Customer lifetime value (CLV): is the total value a customer generates for the company over the course of their relationship with it. For example, at IBM, the CLV could be the money a customer pays the company for the use of their technology services over several years.

CLV = (average revenue per customer) × (number of transactions per year) × (customer lifetime in years)

Where "average revenue per customer" refers to the average revenue generated by an individual customer over a specific period of time, "number of transactions per year" refers to the number of transactions a customer makes in a year, and "customer lifetime in years" refers to the amount of time a customer expects to continue purchasing from the company. The result is expressed in currency.

Of course, there are hundreds and hundreds of metrics, as many as you want to define in your product. Typically, each feature should have an associated quantitative measurement to help you understand the importance it brings to your product.

On the other hand, qualitative metrics are harder to measure and analyze, as they are based on quality and not quantity. Some examples of qualitative metrics include customer satisfaction, customer service quality, and user experience. These metrics are useful for gaining a deeper, more detailed understanding of customer needs and preferences.

Net promoter score (NPS)

It's a measure of customer loyalty to the company. It is calculated by asking customers to what extent they would recommend the company to others. For example, at Microsoft, the NPS could be the result of asking product users if they would recommend Microsoft to their friends or family.

The formula for calculating the NPS is:

NPS = (% of promoters) - (% of detractors)

Where "% of promoters" refers to the percentage of customers who respond with a 9 or 10 and "% of detractors" refers to the percentage that gives a score of 0 to 6. The result is a number between -100 and 100. Higher scores indicate greater customer satisfaction and loyalty, while lower scores indicate the opposite.

Customer satisfaction (CSAT)

It's a measure of customer satisfaction with the company's product or service. For example, at Amazon, the CSAT could be the result of asking customers about their satisfaction with the shopping experience on the site.

The formula for calculating CSAT is:

$$CSAT = (number\ of\ satisfactory\ responses) / (total\ number\ of\ responses) \times 100$$

Where "number of satisfactory responses" refers to the number of customers who respond with a high rating (for example, 4 or 5), and "total number of responses" alludes to the total number of customers who respond to the survey. The result is a percentage. A high CSAT score indicates high customer satisfaction, while a low score indicates the opposite.

User experience or customer behavior: it is a measure of the user's experience with the company's product or service. For example, at Netflix, the UX might be how easily a user finds and plays content on the platform.

Analyzing user experience (UX) is a complex process that involves collecting and analyzing data on how users interact with a product or service. There are several techniques and tools that can be used to analyze the UX, some of which are:

- User satisfaction surveys: users are asked to respond to questions about their experience with the product or service.

- User testing: they are asked to perform specific tasks using the product or service, while an observer records their actions and comments.

- Usage analysis: data is collected on how users interact with the product or service, such as the time they spend on certain pages or the buttons they press.

- User interviews: in-depth interviews are conducted with users to gain a more detailed understanding of their experiences and needs.

- Sentiment analysis: the opinions and comments of users on social media, forums, and other media are analyzed to get a general idea of their experiences.

THE VALUE OF QUALITATIVE METRICS OVER QUANTITATIVE ONES

Something I have often reiterated throughout my career is that the talent of product owners and product managers is seen in the interpretation of metrics, and indeed qualitative metrics are the hardest to interpret, as they are contaminated by a multitude of different contexts. It is here where experience, talent, intuition, and a lot of hard work make a difference between products.

SOME EXAMPLES OF HOW SOME COMPANIES USE QUALITATIVE METRICS TO IMPROVE THEIR PRODUCTS

Amazon uses qualitative metrics, such as customer reviews, which allow it to understand the needs and problems of its buyers more deeply and focus on solving them.

IBM uses interviews and focus groups with its customers to gain valuable information on how they use their products and what problems they encounter. This qualitative feedback has been crucial in improving the quality of their products and adapting to the needs of their customers.

Microsoft uses techniques such as churn rate analysis and product usage tracking to understand how customers interact with their products. This has allowed it to identify problems and opportunities for improvement that could otherwise have gone unnoticed.

Netflix uses qualitative metrics, such as customer satisfaction and loyalty, to understand how their content is performing and how they can improve the user experience. This has allowed it to adapt to changes in the tastes and preferences of its customers and remain a market leader.

Qualitative metrics are especially valuable because they allow us to gain a deeper understanding of the needs and problems of our customers and provide us with the opportunity to adapt and improve accordingly.

Bear in mind that both quantitative and qualitative metrics are important and should be used complementarily to get a complete view of how our product is performing.

VANITY METRICS AND SUCCESS METRICS

Vanity metrics are those that are usually easy to measure and tend to generate a sense of success, but do not necessarily indicate the real success of a product or campaign. For example, the number of followers or visits to a website can be a vanity metric, as it can generate a sense of popularity, but does not necessarily indicate that the product is being successful in terms of generating revenue or meeting customer needs.

On the other hand, success metrics are those that truly indicate the performance and success of a product or a campaign. For example, the number of active customers or the income generated by the product are success metrics, as they indicate that the product is managing to generate income and meet customer needs.

It is important not to get confused with vanity metrics and focus on success metrics because they truly indicate the performance of a product or campaign.

As I have mentioned, every feature, every button, and every behavior of your product must have an associated metric that is clear enough to understand its performance. Usually, these types of metrics should be compared and always viewed alongside associated conversion metrics.

For example, if we need to measure the search functionality of an e-commerce product, we could use the following list of metrics:

Search button click rate

The formula for the search button click rate can be calculated as the number of clicks on the search button divided by the total number of visits to the page and multiplied by 100 to get a percentage.

Search button click rate = (Number of clicks on the search button / Total number of visits to the page) * 100

Search abandonment rate

The search abandonment rate can be calculated as the number of users who started a search on the page and abandoned before finding what they were looking for, divided by the total number of searches initiated and multiplied by 100 to get a percentage.

Search abandonment rate = (Number of abandoned searches / Total number of searches initiated) * 100

Search success rate

The search success rate can be calculated as the number of users who found what they were looking for in a search, divided by the total number of searches initiated and multiplied by 100 to get a percentage.

Search success rate = (Number of successful searches / Total number of searches initiated) * 100

Search conversion rate

he search conversion rate can be calculated as the number of users who completed a desired action, for example in the case of e-commerce having made the purchase of the product found through the search, divided by the total number of searches initiated and multiplied by 100 to get a percentage.

Search conversion rate = (Number of conversions after search / Total number of searches initiated) * 100

For the functionality of the shopping cart:

Cart abandonment rate

The cart abandonment rate can be calculated as the number of users who added products to their shopping cart and then abandoned without completing the purchase, divided by the total number of carts initiated and multiplied by 100 to get a percentage.

Cart abandonment rate = (Number of abandoned carts / Total number of carts initiated) * 100

Cart conversion rate

The cart conversion rate can be calculated as the number of users who completed a purchase after adding products to their shopping cart, divided by the total number of carts initiated and multiplied by 100 to get a percentage.

Cart conversion rate = (Number of completed purchases / Total number of carts initiated) * 100

Average number of products in the cart

The average number of products in the cart can be calculated as the total number of products added to the shopping carts divided by the total number of carts initiated.

Average number of products in the cart = Total number of products added / Total number of carts initiated

Average cart value

he average cart value can be calculated as the total value of all shopping carts divided by the total number of carts initiated.

Average cart value = Total cart value / Total number of carts initiated

HOW TO USE METRICS AND DATA TO MEASURE THE SUCCESS OF YOUR PRODUCT?

Though launching a product is an exciting time, it is crucial to remember that the process of product management does not end once the launch is done. Instead, it is important to keep an eye on the outcomes and the market needs, and to be ready to make adjustments accordingly.

Use Adoption Metrics

These measure the number of users utilizing the product and the frequency with which they use it. Some common adoption metrics include the number of unique users, the number of user sessions, and the average time a user spends using the product. These metrics can help measure product success in terms of its appeal and usability for customers.

Use Performance Metrics

These measure the product's performance and how it compares to business objectives. Some common performance metrics include the Return on Investment (ROI), the Customer Acquisition Cost (CAC), and the payback period. These metrics can help measure the product's success in terms of its contribution to the business and its profitability.

Use Customer Satisfaction Metrics

These measure the customer's level of satisfaction with the product and how it compares to the company's objectives. Some common customer satisfaction metrics are the Customer Satisfaction Index (CSAT) and the Customer Loyalty Index (CLV). These metrics can help measure the product's success in terms of its value to the customer and its potential to retain them.

Adjust the Product

If the results show that the product is not meeting market or customer expectations, it may be necessary to make adjustments to the product to improve its performance and value to the customer. These adjustments can include adding new features, improving existing features, or removing little-used features.

Adjust the Marketing Strategy

If the marketing strategy is not generating the desired impact, it may be necessary to make adjustments to improve its effectiveness. These could be changes in the marketing message, in market segmentation, or in the distribution channel used..

Adjust the Pricing Strategy

If the results show that the product is not generating the desired income, it may be necessary to make adjustments to the pricing strategy. For example, changes in the product's price, in the discounts offered, or in the pricing structure used.

Adjust the Product Management Approach

These adjustments can include changes in the way product objectives are set, in the way product features are prioritized, or in the way the product team is involved in the decision-making process.

Making adjustments to the product, the marketing strategy, and the product management approach is a continuous process, and it is necessary to be attentive to outcomes and market needs to make the necessary adjustments. By making strategic adjustments based on data and metrics, you can improve product performance and maximize its value to the customer.

I have not wanted to emphasize the tools you need because it is very specific to each company, but the most common ones to measure a product's performance are Google Analytics, Mixpanel, Amplitude, Segment, Tableau, and Periscope, among many others. These tools allow you to collect and analyze data about website traffic, user behavior, and key conversion metrics.

There are also specialized tools to measure the performance of specific features, like live chat or customer satisfaction surveys. In addition, many companies use error tracking and performance monitoring tools to measure product behavior in real-time and quickly detect issues..

EXERCISE 8: « GIANT STEPS »

If you have completed the exercises, you should have a results-oriented roadmap based on your OKRs. In this exercise, you should try to create at least three qualitative and five quantitative metrics to help you identify your product's performance. On the next page, I've created some examples that can serve as a guide.

Netflix

- **Qualitative Metrics:**

  1. Content Quality (based on viewer opinions)
  2. Ease of navigation on the platform
  3. User Interface quality

- **Quantitative Metrics:**

  1. Number of subscribers
  2. Total hours of content viewed
  3. Number of new content added per month
  4. Number of plays for each piece of content
  5. Revenue generated from subscriptions

IBM

- **Qualitative Metrics:**

  1. Customer satisfaction level with customer service
  2. Quality of customer care
  3. Quality of technology solutions

- **Quantitative Metrics:**

  1. Revenue generated from consulting services sales
  2. Number of customers
  3. Number of registered patents
  4. Number of employees
  5. Number of consulting service contracts signed

- **Qualitative Metrics:**

  1. Ease of use of products
  2. Customer service quality
  3. Innovation in products

- **Quantitative Metrics:**

  1. Number of Windows users
  2. Number of app store application downloads
  3. Revenue generated from software license sales
  4. Number of employees
  5. Number of registered patents

Amazon

- **Qualitative Metrics:**

  1. Customer service quality
  2. Speed of product delivery
  3. Quality of online shopping experience

- **Quantitative Metrics:**

  1. Number of online product orders
  2. Revenue generated from online product sales
  3. Number of products in inventory
  4. Number of employees
  5. Number of products added to the catalog per month

Exercise 8

Exercise 8

Haskell

```
main = putStrLn "Hello World"
```

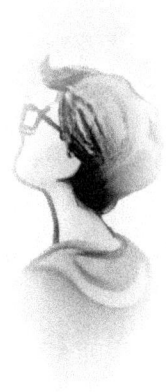

LEVEL 5

CONCLUSION

CREATING EXPERIENCES THAT SOLVE PROBLEMS

From strategy to execution, we are present at every link in the value chain of creating a product or service; what greater satisfaction and reason enough to become product managers!

This is a highly challenging and exciting career that requires a unique set of skills and knowledge. Although the path to product management can be difficult, especially for those with a very technical profile, such as programmers, designers, and other specialists, I believe it has been demonstrated in this book that this transition is possible and that, indeed, it is a natural process in the professional growth of a specialist.

Through the definition of solid strategies and approaches, active customer involvement, and effective collaboration with the development and marketing team, you can become a successful PM and make your products triumph.

In this book, I have tried to highlight the importance of constantly measuring and adjusting the product's success, as well as creating a culture of innovation within the product team.

With these skills and knowledge in your arsenal, you are in an excellent position to become an effective leader.

As you advance your career towards product management, remember the importance of collaborating with the development and marketing team, as well as leading and fostering a culture of innovation. In terms of advice for continuing to progress in this field, I recommend seeking opportunities for continuous learning and keeping up-to-date on the latest trends and tools for your position. It's also important to seek out mentors and role models, either through participating in discussion groups or seeking opportunities that challenge your skills and knowledge.

With dedication and a constant learning attitude, I am sure you will succeed in your career as a Product Manager.

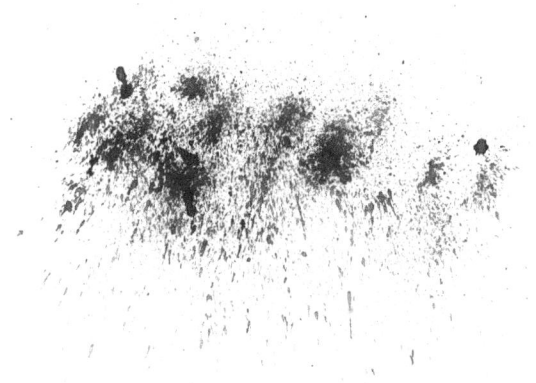

Product management

"Hello User"

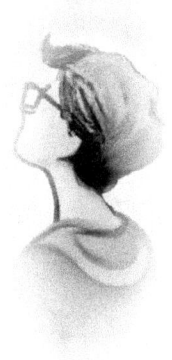

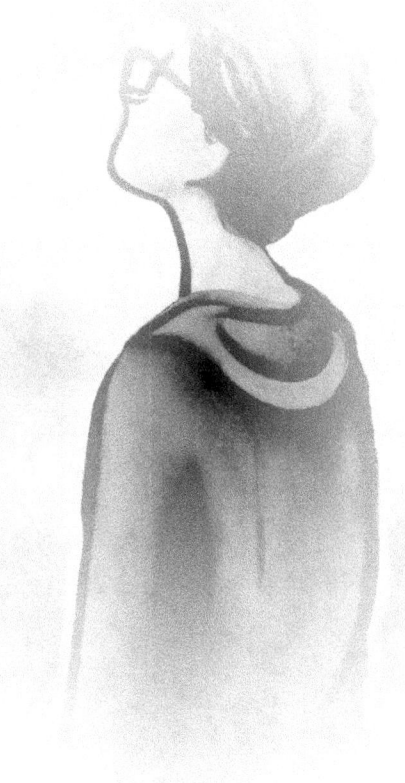

www.ingramcontent.com/pod-product-compliance
Lightning Source LLC
Chambersburg PA
CBHW060826220526
45466CB00003B/989